CONTENTS

POSTERS

MAGNESIUM TECHNOLOGY

PROCEEDINGS
of the conference
sponsored and organized by
the Metals Technology, Metal Science
and Materials Engineering Committees
of The Institute of Metals;
held on 3–4 November 1986
at The Royal Society
in London

The Institute of Metals

1987

Book 396
published in 1987 by
The Institute of Metals
1 Carlton House Terrace
London SW1Y 5DB

and

The Institute of Metals
N American Publications Center
Old Post Road
Brookfield, VT 05036, USA

British Library Cataloguing in Publication Data

```
Magnesium Technology (Conference : 1986 : London).
   Magnesium Technology : proceedings of the conference
   sponsored and organised by the Metal Technology, Metal Science
   and Materials Engineering Committee of The Institute of Metals,
   held on 3-4 November 1986 at The Royal Society in London.
   1. Magnesium
   I. Title   II. Institute of Metals (1985-)
   669'.723     QD181.M4

   ISBN 0-904357-92-9
```

Library of Congress Cataloging in Publication Data

```
Magnesium Technology.
   1. Magnesium--Metallurgy--Congresses.
   2. Magnesium alloys--Congresses.
   3. Magnesium founding--Congresses.
   I. Institute of Metals, Metals Technology Committee.
   II. Institute of Metals, Metal Science Committee.
   III. Institute of Metals, Materials Engineering Committee.
   TN799.M2M26 1987     669'.723     87-3718
   ISBN 0-904357-92-9
```

Compiled by the Institute's CRC Unit
under the editorial guidance of Dr C Baker,
Dr G W Lorimer and Mr W Unsworth
from original typescripts and illustrations
provided by the authors

Printed and made in England by
Whitstable Litho Ltd, Whitstable, Kent

FOREWORD

This Conference was planned to cover the past, present and future of magnesium-based alloys.

The last book to cover magnesium alloys in detail, "Principles of Magnesium Technology" by the late Dr Edward Emley,* was published over 20 years ago. While the present volume can in no way compete with the extensive physical metallurgy coverage in that book, we hope that it will give students and metallurgists in colleges and industry a basis from which to build an understanding of the metallurgy of magnesium alloys and of their advantages and disadvantages compared with competing materials.

Magnesium has long been a "Cinderella" metal to its sister aluminium, its largest market being as an alloying addition for aluminium alloys.

This Conference brought together an international body of speakers to present invited papers covering the world-wide magnesium industry - extraction methods, casting and wrought alloys, and structure and property relationships.

The applications of these products in the aerospace, automotive, nuclear and energy storage fields were presented and so was future technology for overcoming the corrosion problem and for improving processing.

Attendance at the Conference was above expectation and the Organising Committee are indebted to the authors for the high quality of their papers which stimulated great interest and lively discussion.

COLIN BAKER
March 1987

*[Published by Pergamon Press, Oxford, 1966]

CONFERENCE ORGANISING COMMITTEE

DR C BAKER
Alcan International, Banbury

DR G W LORIMER
UMIST, Manchester

MR W UNSWORTH
Magnesium Elektron, Manchester

Alloy and Temper Designations for Magnesium Alloys

The alloy and temper designations for magnesium alloys are given in a four-part code.

The alloy designation consists, first, of letters which identify the major alloying elements, given in order of decreasing concentration.

The letters are followed by numbers which give, to the nearest percentage, the weight percentage of the alloying elements.

The code which relates the letters to alloying elements is as follows:

A	aluminium	N	nickel
B	bismuth	P	lead
C	copper	Q	silver
D	cadmium	R	chromium
E	rare earth	S	silicon
F	iron	T	tin
H	thorium	W	yttrium
K	zirconium	Y	antimony
L	lithium	Z	zinc
M	manganese		

In the ASTM system only the two most concentrated alloying elements are designated. Thus MEL alloy ZCM711, which contains 6.5 wt% Zn - 1.2 wt% Cu - 0.7 wt% Mn has the ASTM designation ZC71.

The third part of the alloy designation consists of a letter which refers to a standard alloy within the broader composition range specified by the first two parts of the designation.

The temper designation, which is separated from the first two or three parts of the code by a hyphen, is the same as that adopted for aluminium alloys. This is given in full in ASTM specification B296-67.

Thus a complete designation might read as

QH21A-T6

The alloy contains approximately 2 wt% silver and 1 wt% thorium; its exact composition is covered by the specification for the "A" version of the alloy; and it is in the solution treated and artificially aged condition. The alloy designation is given in full in ASTM specification B275.

The magnesium outlook

B B CLOW

*The author is Executive Director of
the International Magnesium Association,
Virginia, USA.*

We will briefly review the history of magnesium, production processes that are in commercial use today, location and capacity of the free world's magnesium plants and the markets for magnesium metal. We will conclude by taking a look at the substantial expansion plans that have been recently announced by various producers around the world and the challenge that presents.

HISTORY

While preparing this part of my paper, a brief history of magnesium, I was struck by the coincidence of discoveries and events between magnesium and its sister metal aluminum. I refer to them as sister metals because they are numbers 12 and 13 in the atomic table, they are the lightest and next to lightest of the commonly used structural metals and each is used extensively as an alloying element in the other.

The oxides of both were suspected of having a metallic component in the late 1700's. Antoine Lavosier theorized that "it is highly probable that alumina is the oxide of a metal whose affinity for oxygen is so strong that it cannot be overcome either by carbon or any other known reducing agent".

Sir Humphry Davy had the same suspicions about magnesia, and he confirmed it in 1808 by producing a tiny amount of impure magnesium from a magnesium amalgam which he had made from a moist paste of magnesium and mercury oxides. One year later, in 1809, he identified the metal aluminum by producing an aluminum-iron alloy. Davy suggested the name aluminum for this new metal, a name that is still retained in the United States, but it is referred to as aluminium in most other countries.

The next near coincidence was in 1827 when Freiderich Wohler produced laboratory quantities of aluminum by directly reacting potassium with anhydrous aluminum chloride. One year later, in 1828, A. Bussy did the same thing with anhydrous magnesium chloride.

Next we jump to 1852 when Robert Bunsen built an electrolytic cell that was the forefather of todays cells for electrolyzing anhydrous magnesium chloride. Two years later, in 1854, H. St. Claire Deville built the first commercial aluminum plant that produced the metal by reducing anhydrous aluminum chloride with sodium. The price was high at $115 a pound, but by 1885 they'd worked it down to $11 a pound. Magnesium didn't do much following Bunsen's work in 1852 until the year 1886 which was a banner year for both metals.

In that year, the Hall-Heroult process for aluminum was discovered and it is the basis for all modern day aluminum production.

The first commercial magnesium plant was built by Griesheim Elektron at Stassfurt, Germany, the same year. It used a modified Bunsen cell developed by Fischer and Greatzel and operated on magnesium chloride bittern obtained as a by-product in extracting potash from the carnallite ores at Stassfurt. I. G. Farbenindustrie provided the funds and later took control of the operation. It was later moved to Bitterfeld, Germany, where Pistor and Moschel worked out the process for making virtually anhydrous magnesium chloride in large quantities on a continuous basis.

In 1916, the Dow Chemical Company built its first electrolytic plant to produce magnesium from magnesium chloride bearing well brines at Midland, Michigan. The Dow cell was developed by Herbert H. Dow and his associates to operate on hydrous magnesium chloride cell feed as opposed to the anhydrous requirements of the I. G. Farben process.

Commercial thermic processes for magnesium production didn't get started until World War II. It was during this period that the Amati/Ravelli process of reducing dolomite with ferrosilicon in an internally heated vacuum furnace was developed in Italy. At about the same time the Hansgirg process for the direct reduction of magnesia by carbon was developed in Austria. A little bit later, Dr. Pidgeon and his associates in Canada developed their process for reducing dolomite with ferrosilicon in small, externally heated vacuum retorts. The Hansgirg process was abandoned following World War II, but the other two are still in commercial use. The Magnetherm process, developed by Pechiney-Ugine Kuhlman in the 1950's and 60's, is also a thermic process using ferrosilicon or aluminum as a reductant that operates on a semi-continuous basis with large internally heated furnaces. Over half of the free world's installed thermic reduction capacity is Magnetherm process.

The chlorine gas by-product in the Dow and I. G. Farben electrolytic processes is needed for the cell feed preparation step. It has been a long sought goal in the magnesium industry to produce an anhydrous cell feed by dehydration of magnesium chloride rich brines. This would yield a saleable by-product chlorine that would reduce the costs of the electrolytic process. This was accomplished in

```
┌─────────────────────────────────────────────────────────────────┐
│          FREE WORLD MAGNESIUM PRODUCTION CAPACITY - 1986          │
├─────────────────────────────────────────────────────────────────┤
│                                                                   │
│   Brazil            Silicothermic          10,000  M.T.           │
│   Canada            Silicothermic           9,000  M.T.           │
│   France            Silicothermic          15,000  M.T.           │
│   Italy             Silicothermic          10,000  M.T.           │
│   Japan             Silicothermic (2)      12,000  M.T.           │
│   Norway            Electrolytic           55,000  M.T.           │
│   U.S.A.            Silicothermic          34,000  M.T.           │
│   U.S.A.            Electrolytic (2)      144,000  M.T.           │
│   Yugoslavia        Silicothermic           6,000  M.T.           │
│                                     TOTAL  295,000  M.T.          │
│                                                                   │
└─────────────────────────────────────────────────────────────────┘
```

FIGURE 1.

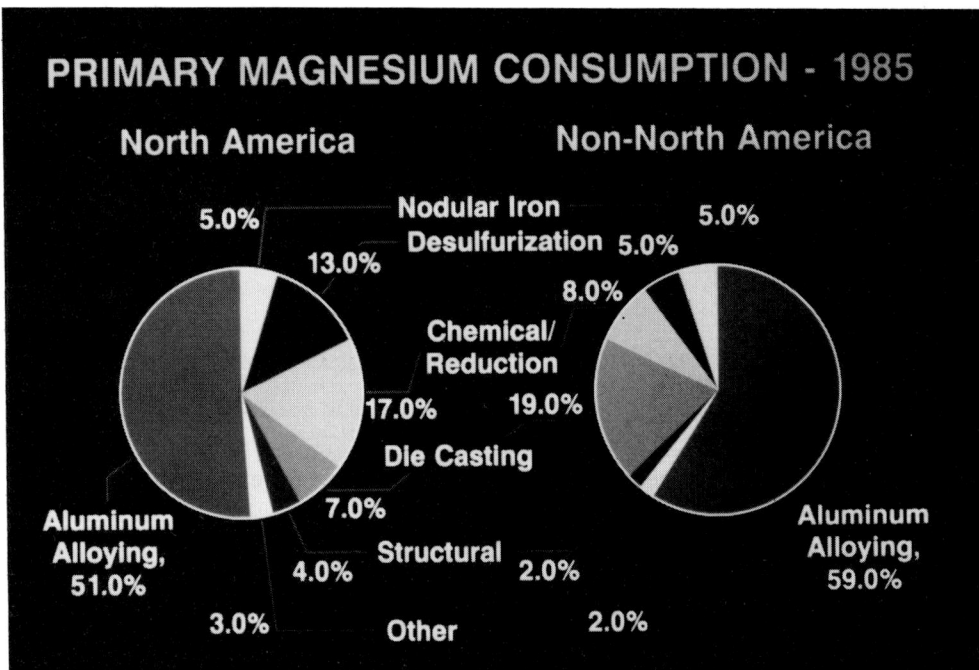

FIGURE 2.

the late 1970's by producers in Europe and North America and should mark an important milestone in the history of magnesium.

PRODUCTION PROCESSES

Magnesium is produced by two principal processes in use today, both being quite well documented in the literature. The processes are the electrolysis of molten magnesium chloride and the thermal reduction of magnesium oxide. Both are used to produce significant quantities of magnesium, although currently the majority of production on a worldwide scale is by the electrolytic method. The magnesium chloride cell feed for the electrolytic process is obtained from the ocean, brines rich in magnesium chloride, residual bitterns from the processing of potash, dolomite or other magnesium bearing minerals. The magnesium oxide furnace feed for the thermal reduction process is obtained from magnesium bearing minerals such as dolomite, brucite or magnesite which are widely distributed in the Earth's crust. The raw material reserves for either process are virtually inexhaustible.

Electrolytic: Three different methods of cell feed preparation for the electrolytic process are in use today. The most difficult part of the cell feed preparation, after eliminating such minor impurities as boron and sulfates, is to separate the magnesium chloride from the six molecules of water that are chemically bound to it. The first $4\frac{1}{2}$ molecules of water can be driven off simply by heating, but the last $1\frac{1}{2}$ molecules will hydrolyze and form undesirable products upon further heating in air. Each of the three methods deals with this problem in a different way.

In the Dow process, magnesium hydroxide is precipitated from sea water by slurrying with calcined dolomite and then converting to magnesium chloride by reacting it with hydrochloric acid. This product is dried to $1\frac{1}{2}$ moles of water and introduced directly into the cells. The by-product chlorine from the electrolytic cells is recycled into hydrochloric acid for use at the front end of the process. Overall, the process is a net consumer of chlorine. The Dow process was first used in 1916 in Midland, Michigan.

In the I. G. Farben process, magnesium hydroxide is calcined to the oxide and chlorinated in the presence of carbon in a verticle furnace to yield a truly anhydrous cell feed. The by-product chlorine from this process is recycled to the chlorinator. This process is also a net consumer of chlorine. It was developed between the years 1926 and 1929 at the I. G. Farben works in Bitterfeld, Germany.

The most recent technology is called the anhydrous process, and starts with a magnesium chloride rich brine which is purified and concentrated to $1\frac{1}{2}$ moles of water by conventional methods. The final dehydration is carried out by proprietary methods to yield a pure anhydrous magnesium chloride cell feed. The by-product chlorine from this process is marketed since it is not needed in the cell feed preparation.

I should mention that a fourth process, developed by Mineral Process Licensing Corporation in England, is in the pilot plant stage with the announced intention to build a 50,000 ton facility using this technology. In this process lump magnesite, which is magnesium carbonate, is continuously fed into the top of a packed bed reactor while chlorine and carbon monoxide gas are injected at the bottom. The process is exothermic at elevated temperature. Anhydrous magnesium chloride is continuously drawn off at the bottom of the reactor and the carbon dioxide exit gas is converted to carbon monoxide for re-use in the process.

Two types of electrolytic cells are in use today; the Dow cell which is an externally heated rectangular steel pot, and the I. G. Farben type cell which is lined with insulating refractory brick contained in a steel tank without provision for external heating.

Thermal: Coincidentally, there are also three variations of the thermic process for reducing magnesium oxide with ferrosilicon that are in commercial use today. The first of these used an internally heated vacuum furnace that was developed in the 1930's by I. G. Farben and separately by Amati in Bolzano, Italy. The I. G. Farben design was discontinued during World War II, but a modern version of the Amati furnace is still being used in Bolzano. More recently, the Brazilians have developed their own model of the internally heated vacuum furnace and are in the process of substantially expanding their facility.

Dr. L. M. Pidgeon and his associates in Canada developed the process that bears his name during WWII. It employs a small diameter, externally heated vacuum retort and produces a very high purity magnesium. A number of these plants were built during World War II, but the only ones operating today are in Canada and Japan.

The Magnetherm process, developed by Pechiney-Ugine Kuhlman in the 1950's in France, employs an electric furnace with a liquid slag. Alumina is added to the charge to reduce the melting point of the calcium silicate reaction product and the calcium aluminum silicate slag is tapped as a liquid. The slag has sufficient resistivity to maintain it in the liquid state by resistance heating. This process is currently used in France, USA and Yugoslavia.

FREE WORLD CAPACITY

Figure 1 shows presently installed free world capacity for magnesium production by country and process. We have not included any of the electrolytic cell capacity installed in Kroll process titanium plants for recycling their by-product magnesium chloride, because this metal is generally not available to the market. You will note that eight of the eleven plants are silicothermic, but 67 per cent of the capacity is electrolytic. It is interesting to note that recently announced expansion plans include two electrolytic plants and two silicothermic plants, but more on this later. The two processes seem to be cost competitive.

MAGNESIUM MARKETS

Figure 2 shows the percentage breakdown of magnesium consumption in 1985. We have divided it between North America and non-North America because the North American market, largely the United States, accounted for almost half of the 225,000 M.T. demand. This has been a consistent pattern for a number of years.

Aluminum alloying is the largest single application for magnesium, accounting for over half of total consumption in both cases. Relatively small additions of magnesium to aluminum will improve its strength and corrosion resistance. Almost three fourths of the registered aluminum alloys contain some magnesium, but the two most

3

```
┌─────────────────────────────────────────────────────────────┐
│                  MAGNESIUM EXPANSION PLANS                   │
│                         1986 - 1992                         │
├─────────────────────────────────────────────────────────────┤
│  ANNOUNCED                                                   │
│  ALCOA/MPLC          CANADA              50,000 M.T.        │
│  CHROMASCO              "                 5,000 M.T.        │
│  NORSK HYDRO            "                60,000 M.T.        │
│  BRASMAG             BRAZIL              25,000 M.T.        │
│  MAGNOHROM           YUGOSLAVIA           3,000 M.T.        │
│                             Sub-total   143,000 M.T.        │
├─────────────────────────────────────────────────────────────┤
│  POTENTIAL                                                   │
│  TAMIL NADU          INDIA               1,000 M.T.        │
│  JAPAN METAL         JAPAN               6,000 M.T.        │
│  ELKEM/BRASMAG       NORWAY             15,000 M.T.        │
│                             Sub-total    22,000 M.T.        │
├─────────────────────────────────────────────────────────────┤
│                          Grand total    165,000 M.T.        │
├─────────────────────────────────────────────────────────────┤
│                                                             │
│             1986 INSTALLED CAPACITY     295,000 M.T.        │
│                                                             │
└─────────────────────────────────────────────────────────────┘
```

FIGURE 3.

PRIMARY MAGNESIUM CONSUMPTION Thousand Metric Tons		
END USE CATEGORIES	PROJECTED 1986	1990
Aluminum Alloying	128	132
Nodular Iron	11	13
Desulfurization	23	33
Chemical/Reduction	27	27
Pressure Die Casting	35	51
Structural	8	10
Other	6	6
Total	238	272

FIGURE 4.

popular ones are alloy 3004 (1.0 to 1.5 per cent magnesium) which is used for beverage can bodies, and alloy 5182 (4.0 to 5.0 per cent magnesium) which is used for the lids. Over 50 per cent of the aluminum beverage cans sold in the United States today are being recycled, thereby conserving several thousand tons of magnesium each year.

One of the fastest growing markets for magnesium is the desulfurization of iron and steel. Sulfur has a deleterious effect on the properties of steel and the increasing demands of the market have forced the steel industry to provide products with lower and lower sulfur content. Magnesium has a high affinity for sulfur and when injected into molten iron or steel, it will reduce the sulfur content dramatically.

Magnesium is also an important element in the production of nodular cast iron which is also known as ductile iron and spheroidal graphite iron. It is produced by introducing a small but definite amount of magnesium and/or rare earths to the molten iron. This causes the graphite in the iron to nucleate as spherical particles instead of flakes, thereby imparting higher strength and much greater ductility to the casting. Ductile iron castings are widely used in automotive components.

There are a number of end uses grouped under the chemical and reduction markets. One of the major ones is magnesium anodes which are used to prevent galvanic corrosion of steel in certain environments such as underground storage tanks, pipelines and domestic water heaters. Also included in this category is its use as a reducing agent in the production of beryllium, titanium, zirconium, hafnium and uranium. Finely divided magnesium is used in pyrotechnics, either as pure magnesium or alloyed with aluminum. It is also used in dry cell and emergency batteries, principally in military applications. In the printing industry, magnesium alloy plates are used for photoengraving because it etches rapidly to provide a sharp impression and the by-products are less hazardous than alternative materials. And last in this category is magnesium tooling plate which is used for making jigs and fixtures because of its high dimensional stability and ease of machining.

The end uses that we have mentioned thus far – aluminum alloying, desulfurization, nodular iron and chemical/reduction – account for roughly 80 per cent of the total demand. A common denominator for these markets is that the demand for magnesium is largely controlled by factors outside of our industry and no amount of selling by the magnesium producers can change the total demand for aluminum alloying. An exception in this group is desulfurization where magnesium competes directly with calcium carbide.

The remaining markets – die casting, structural and other – are what we generically refer to as the structural markets. These are the markets where the magnesium industry can change the total demand by planned market development and aggressive selling.

One of the fastest growing structural markets is die castings, particularly since the intro-duction of the corrosion resistant high purity alloys in the early 1980's. There is a wide diversity of non-automotive die cast parts ranging from computer components, luggage frames, lawn mower decks and chain saw housings to baseball bats, fishing reels, ski binding and archery bow handles. Typical of the numerous automotive die castings are cylinder head covers, clutch housings, wheels, air intake grill, air cleaner covers, transaxle cases and carburetors.

Magnesium's low density is especially important for sand cast aerospace applications. Special alloys containing zinc, zirconium, silver, yttrium and rare earths are used for components operating at temperatures up to 300°C for extended periods of time. Typical applications include gear boxes, canopy frames, engine frames, air intakes and speed brakes.

Magnesium is also used in wrought product form such as extrusions, forgings, sheet and plate. Applications for these mill products range from bakery racks, loading ramps, tennis racquet frames and hand trucks to concrete finishing tools, computer printer platens and nuclear fuel element containers.

Magnesium is a very versatile metal, but like every other metal, and for that matter, each of us, it is not perfect. Its high position in the electromotive force series of the elements is very unforgiving of a design that sets up a galvanic couple. Its close packed hexagonal crystal structure makes it difficult if not impossible to roll into thin gauges of sheet and foil; although the rapid solidification technology that is evolving today may cancel that limitation. Unlike its sister aluminum, it does not form an impervious oxide coating when exposed to the atmosphere, although the importance of this has been significantly minimized since the introduction of the high purity and yttrium alloys. One has to conclude that with good design, and adequate coatings where needed, magnesium parts will give a long and useful service life.

INDUSTRY EXPANSION PLANS

The magnesium industry's expansion plans between now and 1992 are substantial (Figure 3). Five producers have announced that they will build new capacity totalling 143,000 M.T. between now and 1992. In fact, one of these, Magnohrom in Yugoslavia has already completed their 3,000 M.T. expansion. If the other four producers complete their announced projects, 48 per cent will have been added to the existing capacity base of 295,000 M.T. per year. Additionally, three new entrants have stated their intention to build 22,000 M.T. of new capacity subject to certain conditions. If all these projects are completed, it will add 56 per cent to the present free world capacity base. This is the largest and most rapid expansion since World War II, and presents a marketing challenge to the producers.

The IMA surveyed the free worlds' eleven primary producers in early 1986 asking for their forecasts of consumption in 1986 and 1990 (Figure 4). Their forecast for 1986 was 238,000 M.T. which is 80 per cent of their installed capacities. If we project their 1990 forecast to 1992 we get consumption of 286,000 M.T., or an operating rate of 65 per cent if we assume that only the announced capacity increases of 143,000 M.T. will be completed by then. To maintain an 80 per cent operating rate of the 1992 installed capacity of 438,000 M.T., it will be necessary to find a market for an additional 66,000 M.T. – over and above their forecast – during the next six years. This works out to be less than four pounds per vehicle of the free world's automotive production of 40,000,000 units per year. A mere drop in the bucket compared to aluminum usage of over 100 pounds per vehicle. Volvo used over 100 pounds of magnesium in their LCP 2000 prototype vehicle which had a total weight of about 1500 pounds, just half of today's average passenger car weight.

Aggressive market development coupled with the solid advances that have been made in high purity alloys and die casting technology should make this achievable. The producers have demonstrated their faith in the future of magnesium by putting their money on the line, and I have no doubt that they will make it happen because magnesium is on the move.

Extractive metallurgy of magnesium

A M CAMERON, A van HATTEM
and V G AURICH

*Dr Cameron is in the Department of Metallurgy,
UMIST, Manchester, England. Dr Aurich and
Ir. van·Hattem are with Billiton Research BV,
Arnhem, The Netherlands.*

SYNOPSIS

Magnesium is the eighth most abundant element and
the lightest of the constructional metals. Its
production is however energy intensive and these
factors, taken together, provide the motivation
for the improvement of existing, or development of
new, extraction processes.

This paper reviews recent advances in
magnesium production technology and describes new
processes under development. Since magnesium
competes directly with aluminium for many
applications its usage is strongly dependent on
the relative economics of the extraction
technologies. Consequently those developments
offering the greatest potential for a reduction of
the magnesium:aluminium price ratio are
highlighted.

INTRODUCTION

Magnesium and its alloys are characterised by
excellent machinability and good hot formability.
This, in conjunction with the high strength to
weight ratios of castings and stiffness to weight
ratios of the wrought alloys, makes magnesium an
attractive choice for many engineering
applications. Nevertheless magnesium production
levels remain relatively low at approximately
300,000 tonnes per annum. This represents a mere
2% of primary aluminium production (~14m tonnes).
An important factor is the prohibitive Mg:Al price
ratio of 3:1 (July 86 prices - Mg £2.30 per kg -
Al £0.75 per kg). It is claimed that a reduction
of this to a level below 1.6:1 would result in at
least a doubling in usage by the automative
industries, [1]. Potential reductions in
magnesium production costs are therefore of
consequence to future demand.

Magnesium is produced by electrolysis of its
chloride $MgCl_2$ for which important sources are
seawater (0.15 wt% Mg), salt brines such as those
associated with the Great Salt Lake, and
carnallite ($KCl.MgCl_2.6H_2O$). Alternatively
pyrometallurgical processes can be used to reduce
the oxide obtainable from the minerals periclase
(MgO), brucite ($Mg.OH_2$), magnesite ($MgCO_3$) and
dolomite ($MgCO_3.CaCO_3$). Electrolysis is the
predominant route accounting for some 77% of
present day production.

The scope for reducing production costs is
illustrated by comparing theoretical energy
requirements with the actual power consumptions of
existing technologies, Table 1.

Potential for improvement exists in three
distinct but equally critical aspects of the
extractive metallurgy of magnesium. These are,

(i) Manufacture of the reactants. For
electrolysis this generally requires the
production of anhydrous magnesium chloride which
is characterised by high energy requirements
(accounting for approximately 15 kWh/kg.Mg finally
produced). In the production of concentrated
magnesium bitterns from the Great Salt Lake,
yields are as low as 20-30% of total Mg ion
content. Higher overall yields are obtained if
concentrated industrial waste brines are used as
the source of the anhydrous cell feed. Hydrated
feeds can be used but at the expense of reduced
current efficiency in the electrolysis cell.

Metallothermic processes utilise reductants
which are themselves expensive to produce, for
example the Magnetherm process which is the main
competitor to chloride electrolysis, uses 75%
ferrosilicon to reduce calcined dolomite.

(ii) The reduction process. · Here the
objective is the enhancement of space time yields
(kg Mg produced/m^3.hour). Electrolytic winning
processes have some of the lowest processing rates
of the common metallurgical processes, [2]. This
is largely a consequence of the low effective
interfacial areas of standard electrochemical
cells. This can be overcome by the use of novel
reactor designs, such as fluidised bed
electrolysis or bi-polar cells, however many of
these novel reactors are at an early stage of
development and are not yet appropriate to high
temperature molten salt electrowinning or to
processes with liquid products. The simplest
approach to this problem remains the adoption of
higher current densities (insofar as they are not
offset by increased anode overvoltages) and the
minimisation of downtime resulting from the
aggressive environment of the cell.

Productivity of the pyrometallurgical
processes could also be improved if means of
overcoming the present batch-wise operation can be
devised or by making use of high power density
energy sources such as plasma reactors.

(iii) Product Losses: The reactivity of
magnesium can lower yields further due to reaction
with chlorine, oxygen or nitrogen. In order to
minimise the influence many designs of

TABLE 1

COMPARISON OF THEORETICAL AND ACTUAL POWER CONSUMPTION FOR MAGNESIUM PRODUCTION PROCESSES

PROCESS	MAIN REACTION	ΔH_{REACT} (kWh/kg Mg)	POWER CONSUMED BY ELECTROLYSIS OR THERMAL REDUCTION PROCESS (kWh/kg Mg)	OTHER MISCELLANEOUS ENERGY REQUIREMENTS
ELECTROLYSIS	$MgCl_2(\ell) = Mg(\ell) + Cl_2(g)$ 1073K	6.8	12-18	Refining/casting etc. Approximately 15 kWh/kg Mg consumed in preparing anhydrous cell feed.
CARBOTHERMIC REDUCTION	$MgO(s) + C(s) = Mg(g) + CO(g)$ 2100K 298K	8.4	14.4	Strongly dependent on quench choice and efficiency. Sublimation and remelting will take total requirement to >20 kWh/kg Mg.
SILICOTHERMIC REDUCTION	$2MgO(s)+3CaO(s)+1\frac{1}{2}Al_2O_3(s)+Si(s)+Fe(s)$ 298K $= 2Mg(g) + 3CaO.SiO_2.1.5Al_2O_3 + Fe(\ell)$ 1800K	~5.3	17.2	Figure of 17.2 kWh/kg reflects on vacuum operation of reduction process. Ferrosilicon reductant contributes a further 11.7 kWh/kg Mg.

Note: With exception of carbothermic process actual energy requirements are as reported in Proceedings of the International Conference on Energy Conservation in Production and Utilization of Magnesium, M.I.T., 1977. These and additional information from the conference are reproduced in tabular form in reference [16] of main text.

8

electrolysis cell have evolved. Similarly condensor or quench design is the most critical aspect of the pyrometallurgical processes if unacceptable product loss is to be avoided.

The purpose of this paper is to describe recent advances in the extractive metallurgy of magnesium and emphasis is given to developments relevant to the aforementioned key issues.

Magnesium Electrolysis

In recent years few details of the applied technologies have been published. However a survey of the patent literature reveals active research in three main areas, namely, electrolyte composition, cell feed preparation and cell design.

Electrolyte Composition: Attempts to emulate the success of the Hall-Heroult process by development of a technique for the electrolysis of MgO have been unsuccessful. This is primarily due to the limited solubility of MgO in magnesium containing salt mixtures (low melting oxide mixtures are also precluded due to the reactivity of the product magnesium). Recent efforts to formulate oxide containing electrolytes (based on $LiF-MgF_2-NdF_3$ melts) for the production of Mg or Nd have failed as a result of these problems [3].

In spite of these difficulties one of the early processes for magnesium production was based on the electrolysis of an oxide containing fluoride melt. From Harvey's description, [4], it seems likely that an electrolyte composition of 40-50%MgF, 45-50%BaF, 7-10%NaF and 0.1wt%MgO was used.

The reasons suggested for the technical failure of this process are indicative of the characteristics required of a suitable electrolyte; these included
(i) The fluoride melts would not wet the magnesium and therefore afforded little protection against oxidation.
(ii) Slight reduction of the component fluorides lowered the magnesium purity and promoted its oxidation.
(iii) The viscosity of the electrolyte was increased by the presence of suspended MgO particles. Adequate fluidity was only attainable at temperatures of 900-1000°C.
(iv) Magnesium losses resulting from its volatility at the electrolysis temperature were unacceptable.
To date only chloride based electrolytes have been capable of satisfying the requirements of low liquidus temperature, chemical stability, low viscosity and high electrical conductivity. Published electrolyte compositions as quoted by Hoy-Peterson, [5], are reproduced in Table 2. As can be seen most electrolytes are formed from the quaternary system $MgCl_2-CaCl_2-KCl-NaCl$.

Generally it is preferred to have a high electrolyte density relative to that of magnesium (1.58 g cm^{-3} at 943°K). Produced magnesium then floats to the surface of the bath where it can be skimmed-off. In the event of oxidation of the metal the oxides will settle out to the bottom of the cell. High electrolyte density is frequently promoted by selecting an electrolyte composition containing in excess of 30wt% CaCl$_2$ (~2.15 g cm^{-3} at 943°K). Still higher densities are obtainable by means of BaCl$_2$ (~3.25 g cm^{-3} at 943°K) additions which are sometimes used to encourage good phase separation.

The conventional electrolytic cell in which magnesium floats on the electrolyte surface is subject to reduced efficiencies as a result of

recombination of the metal with the evolved chlorine gas. Two approaches have been adopted to solve this problem. The most common solution is to design the cell in such a way as to enable physical separation of the magnesium from the chlorine. An alternative is the use of an electrolyte of low density in order that the magnesium can be tapped off from beneath the protective electrolyte layer. Most 'low density' electrolytes are characterised by a high LiCl$_2$ content, [11], [13] which has the added advantage of considerably increasing the bath's conductivity. This enables higher amperages to be adopted without increasing the cell temperature i.e. improved current efficiencies are attainable. The major disadvantage of these electrolytes is the high cost of LiCl$_2$ which places additional restraints on cell feed purity.

Few systematic studies of the physicochemical properties of salt mixtures suitable for Mg electrolysis have been reported in Western literature. Primary recrystallisation temperatures for ternary and quasi-ternary salt systems formed from LiCl, NaCl, KCl, CaCl$_2$, BaCl$_2$ and MgCl$_2$ have however, recently been determined, [14]. In addition, computer prediction of the phase diagrams of relevant multi-component salt melts is now possible, [15]. Continued improvement of our knowledge of the physico-chemical properties of the chloride electrolytes should result in improved cell efficiencies.

Cell Design: All electrolytic cells for magnesium production are based on one of three designs:
(i) The I.G. cell
(ii) The Dow cell
(iii) Diaphragmless cells
Typical performance data, as reported by Fleming et al, [16] are reproduced in Tables 3-5. The I.G. cell, Fig. 1, consists of four or five carbon anodes contained in an oblong refractory lined steel shell. Each anode is sandwiched between two steel cathodes. Refractory diaphragms are immersed into the electrolyte to effect physical separation of the anode (chlorine) and cathode (magnesium) products. The electrolyte is denser than the magnesium metal which must therefore be collected from a series of small compartments between the electrodes. Temperature is maintained by virtue of resistance heating.

Whilst the use of diaphragms reduces the extent of recombination of the magnesium and chlorine it imposes a large IR drop across the electrolyte by virtue of the large interpolar gap. This results in a severe restriction to the acceptable current density. The cell life of I.G. cells is often dictated by failure of the refractory diaphragms which have to tolerate thermal cycling due to fluctuations in electrolyte level as well as chemical corrosion. I.G. cells are operated with a cell feed of anhydrous magnesium chloride.

The Dow Cell, [17], illustrated in Fig. 2, consists of a steel vessel in a gas-fired brick surround. Cylindrical graphite anodes enter the cell via a refractory cover. Conical cathodes are welded directly to the steel vessel. Since the process utilises a partially dehydrated cell feed corresponding to an average composition of $MgCl_2.1.7H_2O$, [16], [18], heavy anode wear occurs. The hydrated feed has the added consequence that the chlorine gas cannot, unlike the I.G. cell, be recovered as liquefiable Cl$_2$. Since the operation is not dependent on resistance heating of the electrolyte a high conductivity, low density electrolyte (e.g. LiCl/MgCl$_2$) could be chosen.

TABLE 2

Typical Electrolyte Compositions after Höy-Petersen, [5].

% BY WEIGHT						REFERENCE	
$MgCl_2$	$BaCl_2$	$CaCl_2$	KCl	NaCl	LiCl		
15	–	40	25	20		MANTELL	[6]
15	–	40	5	40			
10	–	10	40	40		MUZHZHAVLEV	[7]
10	–	–	75	15		MUZHZHAVLEV	[8]
20	–	–	80	–	–	DOW	[9]
25	–	15	–	60		MANTELL	[6]
15	18	3	20	44		STRELEZ	[10]
13	–	–	–	–	87	DOW	[11]
15	–	20	10	5	50	NATIONAL LEAD	[12]

TABLE 3

Typical Performance Data for an I.G. Cell

Operating temperature	740°C
Cell voltage	5-7 volts
Current	18-150 kA
Current efficiency	80-90%
Energy efficiency	30-35%
Energy consumption	15-18 kWh/kg.Mg
Anode graphite consumption	0.02 kg/kg Mg
Production rate	approx. 200 kg per day

TABLE 4

Typical Performance Data for a Dow Cell

Operating temperature	700°C
Cell voltage	6.0 volts
Current	90 kA
Current efficiency	75-80%
Energy efficiency	30-35%
Energy consumption	18.5 kWh/kg.Mg
Anode graphite consumption	0.1 kg/kg Mg
Production rate	approx. 500 kg per day

The small interpolar gap allows the adoption of high current densities without overheating, although this is constrained by the anode overvoltage.

Diaphragmless cells rely on circulation patterns within the melt to effect separation of product magnesium and chlorine. Many of these cells have been developed in the USSR and Russian cell technology (see [19]) is often regarded as being amongst the most advanced in the world. In the West, Alcan have developed a diaphragmless cell, [20],[21], and it is thought, [16], that Norsk Hydro are using a cell similar to that described in a 1975 patent, [22]. Both of these cells exhibit a main electrolysis or gas separation chamber and a collecting chamber. In the main chamber chlorine is evolved at the anodes and is removed from the cell. The Alcan cell operates at relatively low temperatures (approx. 670-700°C) so as to maintain a relatively viscous electrolyte. Magnesium formed on the cathode then forms a continuous film undisturbed by convective flows in the electrolyte. The metal rises slowly into inverted steel troughs positioned above each cathode, Fig. 3. These troughs divert the metal into the main collection chamber so as to effect the separation from the chlorine.

The Norsk Hydro cell utilises the gas lift/convective flow generated at the anode to force the electrolyte and metal through hollow cathodes situated in the main electrolysis chamber and into a connecting metal separation chamber, Fig.4. It is claimed that separation of gas and metal is efficiently achieved with contact time between the two products reduced to a minimum.

Diaphragmless cells have large (relative to I.G. cells) inter-electrode working volumes. Interpolar gaps are smaller and higher current densities can be expected. The improved current efficiencies attained lead to significantly lower specific energy consumptions than is possible in the traditional cells, Table 4. Included in the table are performance characteristics for a pilot scale diaphragmless cell recently described by Desikan, [23]. It is reported that the low power consumption of this cell reduces the total energy requirements to produce one tonne of 'end product' magnesium from its raw materials to within 6% of the comparable figure for aluminium. If only electrolysis is considered the power consumption of the modern diaphragmless cells (13-15 MWh/tonne Mg) compares favourably with the best of the modern aluminium cells (14 MWh/tonne Al). Furthermore, an important achievement has been the automation of virtually all cell house operations resulting in economic benefits which are of equal importance to those due to reduced power consumption.

It is worth noting that the theoretical decomposition voltage for pure liquid magnesium chloride, according to

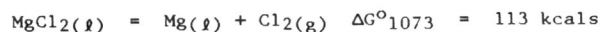

$$MgCl_2(\ell) = Mg(\ell) + Cl_2(g) \quad \Delta G^o{}_{1073} = 113 \text{ kcals}$$

is -2.45V at 800°C. This is significantly lower than the cell voltages of Table 4 suggesting that further reductions in power consumption should be possible by, for example, further reductions in interpolar gap or increasing the activity of $MgCl_2$ in the electrolyte. The extent of any further decreases in power consumption is however dependent on the overall heat balance for the cell.

Cell Feed Preparation: Energy consumption only accounts for some 20% of the total cost of magnesium production. The single most expensive item, reported to account for 47% of total production costs, [16], is the anhydrous $MgCl_2$ feedstock as is required by all cells except Dows.

Two basic technologies have emerged for the primary production of anhydrous $MgCl_2$, ($MgCl_2$ is also formed as a by-product of the Kroll process in which magnesium is used to reduce $TiCl_4$ to titanium metal).

(i) Dehydration of $MgCl_2.xH_2O$ as recovered from aqueous solutions (brines, seawater or industrial waste liquors).

(ii) Direct chlorination of magnesium oxide.

The first of these two processes is conducted in steps involving solar evaporation to obtain concentrated bitterns (approx. 30% $MgCl_2$), purification to remove or reduce sulphate and boron concentrations, dehydration and probably a carbo-chlorination step, [18]. Large quantities of solutions must be handled and magnesium losses are high (70-80% of total Mg ion content). If industrial waste liquors (typically containing 30-35% $MgCl_2$) are available the initial concentration step and associated Mg losses can be eliminated. The main problems are associated with the dehydration normally conducted by spray drying a concentrated solution of the purified chloride. There is a tendency for hydrolysis to occur resulting in the breakdown of magnesium chloride according to

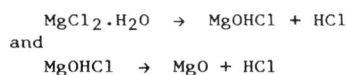

$$MgCl_2.H_2O \rightarrow MgOHCl + HCl$$
and
$$MgOHCl \rightarrow MgO + HCl$$

These reactions can be suppressed by heating in an atmosphere containing dry HCl or by the presence of small amounts of alkali chlorides. Nevertheless the final product of spray drying typically contains 5% MgO and 5% H_2O, necessitating a final carbo-chlorination step. This final chlorination which takes place in e.g. an electrically heated shaft furnace has the advantage of further removing undesirable impurities which may otherwise reduce the efficiency of the electrolysis cell. (The presence of 10 ppm boron in the electrolyte is reported to reduce the current efficiency by approximately 1%.)

An alternative dehydration process route, practiced by Norsk Hydro, involves the drying in fluidised bed or shaft kilns of hydrated magnesium chloride prills formed by centrifuge or sieve-plate spraying of a suitable melt. In this case the final dehydration, conducted under HCl, is repeated until acceptable MgO and H_2O contents are obtained. Other dehydration processes using carnallite ($KCl.MgCl_2.6H_2O$) or ammonium carnallite ($MgCl_2.NH_4Cl.6H_2O$) feedstocks have been reported [19].

Production of anhydrous magnesium chloride by carbo-chlorination of the oxide was a source of cell feed for the early IG-MEL process, [24] and is still practised by Norsk Hydro, [16]. In the original process pelletised feed consisting of MgO, coal, peat and some $MgCl_2$ solution as binder, was charged to an electrically heated shaft furnace into which chlorine gas was injected, Fig. 5. The reaction temperature was approximately 1000°C.

Few details of carbo-chlorination technologies have been reported; however the current preference for dehydration technologies may be a result of one or more of the following factors;

(i) Poor selectivity. Impurities contained in the feedstocks will have been subject to

11

TABLE 5
Power characteristics of various electrolytic cells

Cell Type	Cell Current (kA)	Current Efficiency (%)	Voltage (Volts)	Specific Energy Consumption (kWh/kg Mg)
Alcan	80	90–93	5.7–6	14
Dow	90	75–80	6–7	18–19
IG	150	80	5.5–7	15–18
Soviet Diaphragmless	200	85–90	5–6	13–15
Norsk Hydro	250–350	92–93	4.9	12–13
CECRI (Pilot Plant) [23]	8	71	4.2	13

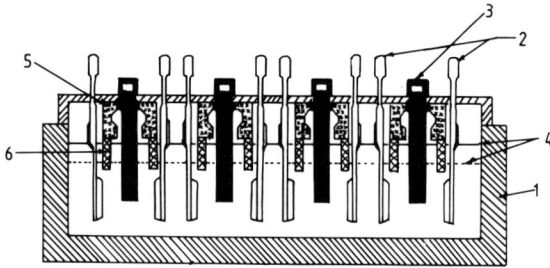

FIG. 1.   I G ELECTROLYSIS CELL

1 – Refractory lining
2 – Steel cathodes
3 – Graphite anode
4 – Upper and lower (dashed line) electrolyte levels
5 – Anode box
6 – Refractory 'semi-walls' or 'diaphragms'.

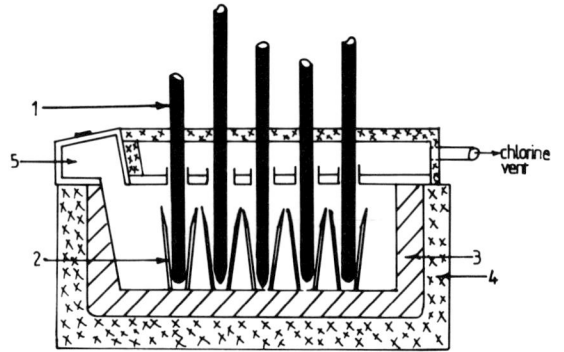

FIG 2.   SCHEMATIC DOW CELL

1 – Cylindrical graphite anodes
2 – Conical steel cathodes
3 – Steel shell
4 – Gas-fired outer shell
5 – Magnesium collection sump.

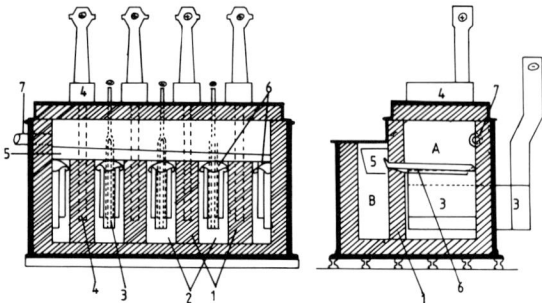

FIG. 3.   SECTIONAL VIEWS OF AN ALCAN CELL

1 – Refractory partition wall

2 – 'Doorways' joining electrolysis (A) and metal separation (B) compartments

3 – Cathodes
4 – Anodes
5 – MG collecting box
6 – Cathode troughs
7 – Gas Vent

A – Electrolysis Zone
B – Metal Separation Zone

carbo-chlorination possibly lowering the purity of the $MgCl_2$.

(ii) Low space/time yield. There does not appear to have been an investigation of the kinetics of the carbo-chlorination reaction. The mechanism (involving two solid and one gaseous reactant) is likely to be complex and could have resulted in an unacceptably slow reaction.

(iii) Oxides present in the ash of the coke may have caused sintering and have led to an uneven descent of the burden. Disposal of the ash will have been problematic and entrainment in the liquid chloride may have adversely influenced product purity.

(iv) Environmental problems are likely due to oxy-chloride and chlorine containing waste gas.

Some of these difficulties may be overcome in an alternative carbo-chlorination process being developed by Minerals Processing Licensing Corporation. In this case two separate reactors are used to effect the reactions;

$$MgCO_3(s) + CO(g) + Cl_2(g) = MgCl_2(\ell) + 2CO_2$$

Reactor I

and

$$2CO_2(g) + 2C(s) = 4CO(g) \qquad \text{Reactor II}$$

Consequently the need to agglomerate the feed is removed and it is also conceivable that enhanced reaction kinetics are obtained. By eliminating the solid reductant and its associated impurities from the chlorination reactor and by making use of high purity magnesite a purer product is likely. Initial pilot plant trials have been completed and it is claimed that successful implementation of this technology could lead to a very significant reduction in magnesium production costs. An assessment of the validity of this claim is prevented by the absence of published data.

Thermic Processes

The main attraction of pyrometallurgical routes to magnesium producers is the ready availability of suitable feedstocks. In comparison to the electrolytic process a simple pretreatment of calcination and perhaps agglomeration can be adopted. Unfortunately the chemical stability of MgO requires that the reduction be effected at high temperature or low pressure or a combination of both. In the case of metallothermic reduction the cost of the reductant is a significant economic penalty.

Metallothermic Reduction: Al, Ca, Si, their alloys or carbides are all, under appropriate conditions of temperature and pressure, capable of reducing magnesium oxide. However, for what are essentially economic reasons, (see [16]), only ferro-silicon has emerged as a viable metallic reductant.

The Pidgeon Process: [25] This was one of the earliest and most successful silicothermic processes. Briquettes consisting of ferro-silicon and calcined dolomite are charged in 150 kg batches to tubular steel retorts. The retorts are externally heated to a temperature of approximately 1200°C in gas fired or electrical furnaces. To enable reaction to occur a vacuum of 0.1 torr is maintained. Produced magnesium collects in the form of 'crowns' within a water cooled section of the retort located outside the main furnace, Fig. 6.

A major deficiency of the Pidgeon process is the lengthy cycle time of approximately 8 hours which is required to produce some 20 kg of magnesium per retort. Processing rates cannot be improved by raising the temperature without risking damage to the retorts. The efficiency with which the silicon is utilised is poor since reaction of low activity silicon (a_{Si} falls rapidly to less than 0.1 below the stoichiometric composition FeSi) would require still longer cycle times. Finally collection of the magnesium in the form of solid 'crowns' necessitates remelting of the product.

The Magnetherm Process: This has superceded the Pidgeon process as the main silicothermic production route and is, at present, the main competitor to chloride electrolysis.

The process involves the reduction of magnesia, introduced to an electric arc furnace in the form of calcined dolomite, by ferro-silicon. In this sense the process is directly comparable to the Pidgeon process; however the space time yield is considerably higher due to the reaction being in the liquid phase and at higher temperature (1550°C). Liquid phase reaction is accomplished by fluxing the dicalcium silicate reaction product with a carefully controlled quantity of alumina or bauxite so that the overall reaction becomes

$$2CaO.MgO + (xFe)Si + nAl_2O_3 \rightarrow$$

$$2CaO.SiO_2.nAl_2O_3 + 2Mg + xFe.$$

The reaction is promoted by the low silica activity in the resultant slag and by operation under a vacuum of 0.05 atm, [26],[27]. The slag composition of 55%CaO, 25%SiO$_2$, 14%Al$_2$O$_3$ and 6%MgO has an estimated liquidus temperature of 1800°C. As a result an equilibrium is established at the operating temperature of 1550°C between solid 2CaO.SiO$_2$ and a liquid component with an estimated composition of 48.3%CaO, 18.3%SiO$_2$, 23.3%Al$_2$O$_3$ and 10%MgO. This latter composition is situated on the phase boundary between the primary recrystallisation fields of dicalcium silicate and periclase and is therefore doubly saturated with respect to 2CaO.SiO$_2$ and MgO. Since the two phase oxide mixture will contain some 40% solids accurate composition and temperature control must be required to prevent stalling of the slag and to enable tapping.

With the slag composition and therefore silica and magnesia activities defined as outlined above the pressure of magnesium generated becomes a function of temperature and silicon activity only. It is easily shown that, provided equilibrium is attained, the residual ferro-silicon will have a silicon activity of approximately 0.01 which corresponds to about 20% Si. Higher magnesium pressures are only obtainable by accepting still lower efficiencies of silicon utilisation.

The need for low pressure operation results in the main disadvantage of the Magnetherm process since batchwise operation is necessitated with a daily downtime in excess of 15%. Furthermore ingress of air to the condenser causes formation of Mg$_3$N$_2$ and MgO which accounts for some 20% product losses. A simplified flowsheet for the process is given in Fig. 7.

Carbothermic Reduction: All pyrometallurgical processes for magnesium production generate gaseous products from which the magnesium must be recovered. The carbothermic processes have the additional constraint that the high temperature gas-phase reaction between magnesium and the

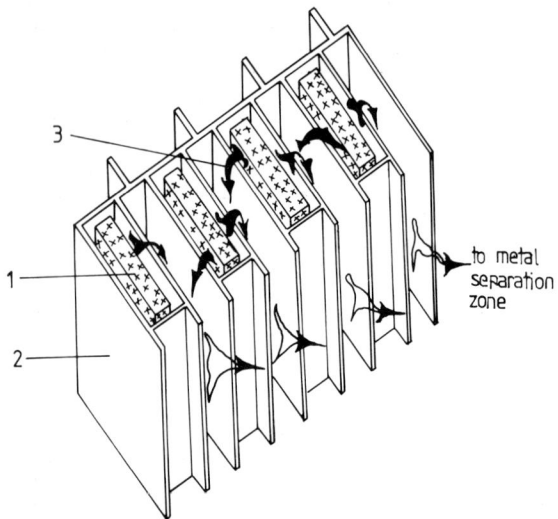

FIG. 4. ELECTRODE ARRANGEMENT AND ELECTROLYTE
FLOW PATTERN IN A NORSK-HYDRO CELL

1 - Anode blocks
2 - Steel cathode
3 - Electrolyte flow

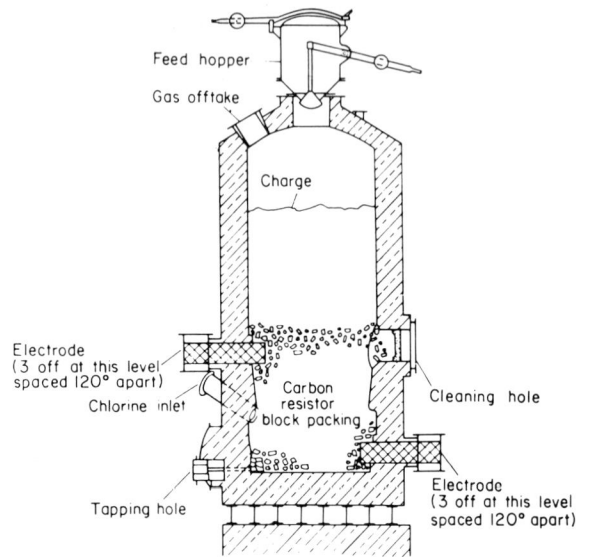

Fig. 5. Chlorinator for production of anhydrous
magnesium chloride. Reprinted with permission
from E.F. Emley, "Principles of Magnesium
Technology", Copyright 1966, Pergamon Books Ltd.

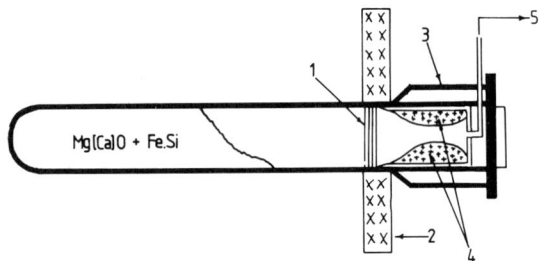

FIG. 6. PIDGEON PROCESS RETORT

1 - Radiation baffles
2 - Furnace wall
3 - Water cooled condensor
4 - MG 'crowns'
5 - To vacuum pumps.

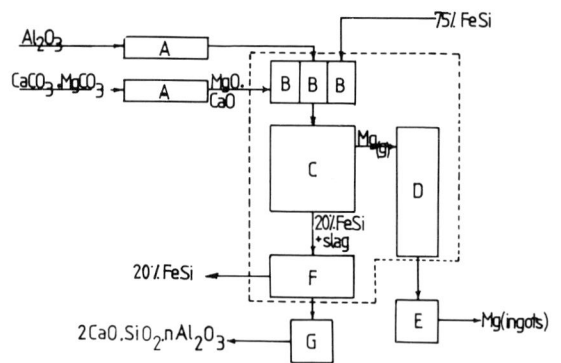

FIG. 7. FLOW SHEET FOR MAGNETHERM PROCESS

A. Rotary kilns
B. Feed bins
C. Arc furnace
D. MG condensor
E. MG foundry
F. Slag/metal tapping and
    settling ladles
G. Slag granulator

-   units enclosed by dashed box operate
    under vacuum.

co-produced carbon monoxide will occur (under one atmosphere total pressure) at temperatures below 1780°C. Either rapid quenching (<20 ms, [30]) or separation of the gaseous products must be effected to prevent unacceptable magnesium losses. In spite of this, development of a successful carbothermic technology is very attractive due to the potential energy efficiency associated with the direct use of carbon as reductant.

An early carbothermic process was developed by Hansgirg, [28], [29] and operated at Permanente [30]. A flowsheet for the process is given in Fig. 8. The reduction reaction was carried out on top of a metallurgical coke bed, heated in an arc-furnace to temperatures of about 2000°C. The pressure inside the reactor was slightly greater than atmospheric. Product gases were shock cooled to a temperature below 250°C in a stream of natural gas. Judging from the product composition, the efficiency of this quench step was about 80%. Energy consumption for the reactor and quench section was about 12.4×10^6 kcals/ton Mg, (14.4 kWh/kg) [31]. Total energy requirements will have been much higher as a consequence of sublimation and remelting operations. A serious deficiency was the formation of pyrophoric magnesium dust during quenching which rendered the process dangerous.

Recently Billiton Research B.V. of The Netherlands have been developing an improved version of the carbothermic process, [32]. Pelletised magnesia and coke is charged to an electric arc furnace containing a slag with nominal composition 20wt% MgO, 35%CaO and 45%Al_2O_3 which has a liquidus temperature of approximately 1800°C. As the slag composition lies in the periclase region of the phase diagram the magnesia activity is close to unity at the operating temperature of 1800-1850°C. Resistance heating of the slag is employed and the slag then acts as both a heat transfer medium and a sink for feedstock impurities. Up to 15wt%SiO_2 can be accommodated without any lowering of the magnesium purity which is typically in excess of 99% Mg, Fig. 9. Plasma arc heating can also be used and the high temperatures and reaction rates generated enable the use of non-agglomerated feedstocks. Indeed it has been demonstrated that non agglomerates improve magnesium purity levels when plasma heating is used. This is thought to be due to reduction of the dissolved magnesia from the surface layers being the predominant reaction mechanism under these conditions. In other words co-reduction of agglomerated impurities is eliminated. This concurs with the observations of Mintek on the plasma processing of ferro-alloy oxides,[33].

Development of techniques to obtain the high heat and mass transfer rates required for an improved quench step is ongoing. For example, many of the difficulties associated with the quench step could be overcome by quenching directly to a liquid by means of liquid metal spraying.

DISCUSSION AND CONCLUSIONS

The introduction of diaphragmless cells, cell house automation and the development of the Magnetherm process represents major advances with respect to the early electro and pyro-metallurgy of magnesium. Nevertheless there remains significant scope for improvement which, in conjunction with the attractive physical properties of the metal, represents a considerable opportunity for growth of the magnesium market.

New or improved technologies for cell feed preparation can reasonably be expected. This may result from a better understanding of magnesium chloride dehydration technologies; for example, an understanding of the influence of alkali chlorides on the relative rates of hydrolysis and dehydration could pay dividends. Alternatively the successful development of new environmentally acceptable carbo-chlorination techniques would be a major factor in reducing production costs.

Further advances in cell technology are equally feasible. Improvements in diaphragmless cells could result from a better understanding of electrolyte circulation patterns. Mathematical and physical modelling would assist particularly if sufficient data on physicochemical properties such as viscosity, density, conductivities and surface tension were made available. In this context it is interesting to note that a method for electrolytic production in which very small density differences are utilised has recently been patented, [34]. It is claimed that separation of the magnesium from the electrolyte can be controlled so as to take place entirely in the separation chamber, resulting in reduced losses due to oxidation or recombination with chlorine. The high electrical conductivity of the $MgCl_2$, $NaCl_2$, $KCl/LiCl$ electrolyte is also claimed to enable higher current densities. Cell power consumptions as low as 9.94 kWh/kg Mg have been demonstrated.

Operating costs for electrolysis are dependent on cell life which is, in turn, related to the choice of refractory, [3]. The use of magnesia alumina spinel ($MgAl_2O_4$) has been suggested as it is claimed that this is more resistant to thermal and chemical degradation than traditional alumina or alumino-silicate refractories, [35].

An interesting approach to the problems of oxide electrolysis is the use of a composite anode consisting of a mixture of carbon and magnesium oxide, [36]. Internal conductors (magnesium rods), dimensioned to recede at a rate approximating the consumption of the anode mixture, allow a near constant anodic current path to be maintained.

The development of novel pyrometallurgical processes is constrained by the availability of economic reductants. Reduction of anhydrous $MgCl_2$ by CaC_2 according to $MgCl_2 + CaC_2 = CaCl_2 + Mg + 2C$ which has recently been researched in France, [37], seems unlikely to be commercially exploited for this reason. The most promising developments are those representing improvements to the silicothermic or carbothermic processes.

The use of liquid metal solvents during carbothermic reduction has been suggested, [38]. The idea is to drastically lower the activity of the produced magnesium by utilising the large negative deviations from ideality associated with some liquid metal solutions (e.g. Mg-Bi, Mg-Sb solutions) according to

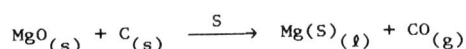

$$MgO_{(s)} + C_{(s)} \xrightarrow{S} Mg(S)_{(\ell)} + CO_{(g)}$$

where S represents the solvent. Such a process is however critically dependent on a high rate of solution of the product magnesium. Furthermore the low activity of the magnesium will render its subsequent recovery difficult. Vacuum distillation has been suggested to effect recovery but (if successful) would be subject to magnesium

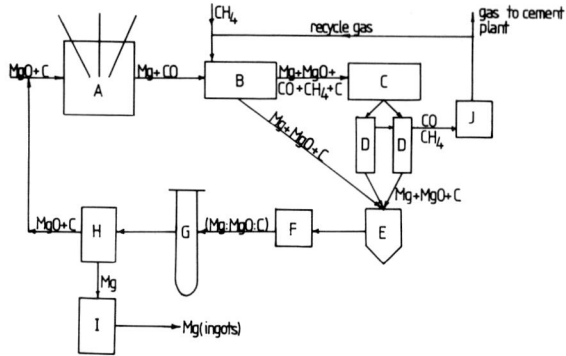

FIG. 8.   FLOW SHEET FOR PERMANENTE PROCESS

A. Arc furnace
B. Quench unit
C. Gas cooler
D. Bag-filters
E. Solids collection bin
F. Pelletiser
G. Retorts
H. MG cooling and stripping
I. MG foundry
J. Gas blower

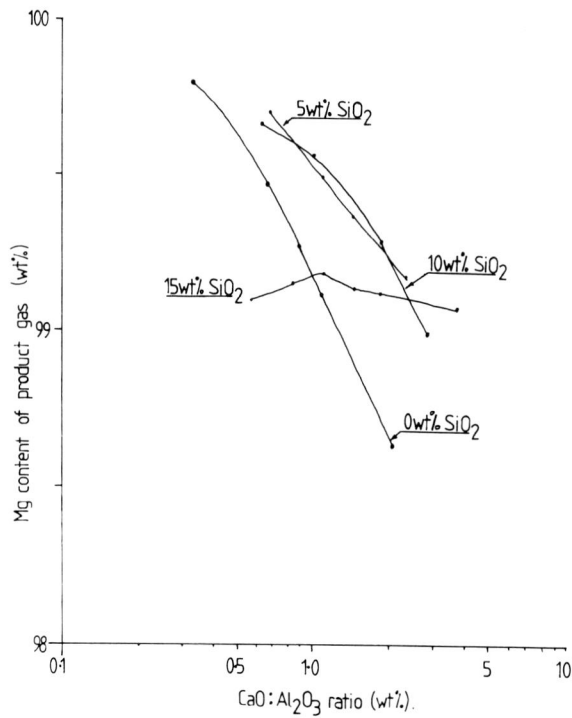

FIG. 9.   MAGNESIUM PURITY VS. SLAG COMPOSITION FOR
BILLITON PROCESS.

16

losses in a manner similar to that experienced in the vacuum condenser of the Magnetherm process. Improvements in condenser (Magnetherm) or quench (Carbothermic) technologies still offer the greatest potential for the pyrometallurgical routes to magnesium.

REFERENCES

[1] Holland, M.S. "Use of Magnesium in Automobiles". Proceedings of the International Conference on Energy Conservation in Production and Utilisation of Magnesium, M.I.T. 1977.

[2] Richardson, F.D. "Interfacial Phenomena and Metallurgical Processes". Canadian Metallurgical Quarterly, vol. 21, No. 2, pp. 111-119, 1982.

[3] Belov, S.F. Avaeva, T.L., Novichkov, V.Kh. and Rybkini, R.D., Izv.V.U.Z. Tsvetn.Met., 1982, (4), 83-87.

[4] Harvey, W.G., Trans.Electrochem.Soc., 1925, 47, 327.

[5] Höy-Peterson, N., Journal of Metals, April 1969, 43-49.

[6] Mantell, C.L., Electrochemical Engineering, McGraw-Hill Book Co., New York, 1960.

[7] Muzhzhalev, K.D. et al., Tsvetn.Metal, vol. 38, (3), 60-65, 1965.

[8] Muzhzhalev, K.D. and Lebedev, O.A., Izv.V.U.Z. Tsvetn.Met., 1960 (5), 89-94.

[9] Dean, L.G. et al., U.S. Patent 2880151, 31.3.1959., (Down Chem.Corp.).

[10] Strelez, C.L. and Bondarenko, N.VB., Tsvetn. Metal., vol. 33, (9), 62-66, 1960.

[11] Williams, E.J. et al., U.S. Patent 2888389, 26.5.1959., (Dow Chem.Corp.).

[12] Love, F.E., U.S. Patent 3389062, 18.6.1968., (National Lead Comp.).

[13] Japanese Patent Publication No Sho 36-9055 1961.

[14] Belvadi, J.B. et al., Advances in Electrometallurgy, Proc.Symp. 1983, Cent. Electrochem.Res.Inst., Karaikudi, India, pp. 2/27-2/43.

[15] Lin, P.L., Pelton, A.D., and Bale, C.W., Journ.Amer.Ceram.Soc., vol. 62, 1979, pp. 414-422

[16] Flemings, M.C. et al., "An Assessment of Magnesium Primary Production Technology". Materials Processing Center, Massachusetts Institute of Technology, Cambridge, U.S.A., 1981. Report to United States Department of Energy.

[17] Schambra, W.P., Trans.Amer.Inst.of Chem.Eng., vol. 41, 1945, pp. 35-54.

[18] Fougner, S. "New Directions in Electrolytic Production of Magnesium". 103rd AIME Annual Meeting 1974, Light Metals Sect. Proc.of Conf., vol. 2, pp. 515-533.

[19] Strelets, Kh.L. "Electrolytic Production of Magnesium", Moscow 1972. English Translation U.S. National Technical Information Service, TT 76-50003, 1977.

[20] Sivilotti, O.G. et al., U.S. Patent 3,396,094. 6th August 1968.

[21] Sivilotti, O.G., Light Metals Age, vol. 42, part 7-8, pp. 16-18, Aug. 84.

[22] Andreassen, K.A. et al., U.S. Patent 3,907,651, 23rd Sept. 1975.

[23] Desikan, P.S., Light Metals: Science and Technology, Proc.Conf., Varanasi, India, 14th-16th Nov. 1983.

[24] Emley, E.F., 'Principles of Magnesium Technology' Pergamon Press, New York, London. 1966.

[25] Froats, A., "Pidgeon silicothermic process in the '70's", Light Metals, New York, 1980, pp. 969-979.

[26] Faure, C. and Marchal, J. "Magnesium by the Magnatherm Process", Journal of Metals, Sept. 1964, pp. 721-723.

[27] Christini, R.A. "Equilibria among Metals; Slag, and Gas Phases in the Magnatherm Process", Light Metals, New York, 1980, pp. 981-995.

[28] Hansgirg, F.J., The Iron Age, 18th Nov. 1943, page 57.

[29] Hansgirg, F.J., The Iron Age, 25th Nov. 1943, page 52.

[30] Dungan, T.A. "Production of Magnesium by the Carbothermic Process at Permanente", Trans AIMME, vol. 159, 1944, pp. 308-314.

[31] Landis, W.S., Trans.Electrochem.Acta, vol. 72, 1937, p. 293.

[32] Warren, G.F. and Cameron, A.M. "Process for Producing Magnesium", European Patent Appl. No 84201741.0, 30th Jan. '85. (Shell Int.Research).

[33] Curr, T.R., Maske, K.U. and Nicol. K.C. "The attainment of high power densities in transferred-arc plasma smelting processes", Symp.Proc.-Int. Symp.Plasma Chem., 7th, vol. 4, 1985, pp. 1186-91.

[34] Ishizuka, H. "Method for Electrolytically obtaining Magnesium Metal" U.S. Patent No. 4,495,037, Jan. 22nd 1985.

[35] Henslee, W.W. et al., European Patent Appl. No. 84100122.2, 8th Jan. 85, (Dow Chem.Corp.)

[36] Upperman, G.V. and Withers, J.C. U.S. Patent No. 4,409,083, 11th Oct. 1983.

[37] Private correspondence.

[38] Eckert, C.A. et al., Ind.Eng.Chem.Process Des Dev., vol. 23, pp. 210-217, 1984.

New developments in casting of magnesium

N ZEUMER, H FUCHS and G BETZ

The authors are all with Honsel-Werke AG, Meschede, Germany.

SYNOPSIS

Although the use of magnesium castings has decreased during recent years, for reasons which will be briefly discussed, a lot of progress has been made as can be seen with the state of the art today. To illustrate this, attention will be given to specific examples. The future prospects of magnesium casting can be improved by applying new developments and technologies - in particular:

(i) new alloys,
(ii) melt handling,
(iii) mould making,
(iv) casting processes and
(v) the application of numerical simulation in casting development.

These will help to produce more cost-efficient castings and stimulate the demand for their use.

PRESENT SITUATION OF MAGNESIUM CASTING

Magnesium and its alloys are still the engineering materials with the lowest densities available on a technical scale. The latter show good mechanical properties at room temperatures (Fig. 1) and remarkable properties at elevated temperatures (Fig. 2), thus making these materials the obvious choice for certain applications.

Unfortunately however, the use of magnesium castings seems to have decreased over the last two decades. The main reason for this appears to be a lack of experience on the part of the potential appliers. Magnesium is usually rated as being highly corrosive and many engineers do not realise that this can be safely treated.

In addition to this the cost of magnesium is higher than that of aluminium, and its casting, machining and surface treatment processes are more expensive as well.

Furthermore the number of foundries casting magnesium components is steadily decreasing, therefore magnesium is not applied as frequently as it used to be.

This seems rather unfortunate when one looks at what has been achieved in magnesium casting to date.

STATE OF THE ART OF MAGNESIUM CASTING

Despite the problematic situation described above, magnesium foundry technology has progressed over recent years and produced many superb castings which make good use of the properties of magnesium. The main casting technology for magnesium is sand casting.

Increasing demands from aerospace industries for more compact and efficient components have been fully met. Wall thicknesses as thin as 3.5mm and casting tolerances of ± 0.5mm can be achieved. Precast lubrication and hydraulic passageways and diameters as narrow as 2mm are now standard practice. These piping systems are able to stand pressures of up to 90 bar.

Examples of the use of highly integrated castings are auxiliary power units for aeroplanes. The gearboxes of these units were amongst the first where this technology has been applied. Their housings and their lids incorporate a number of precast passageways. Fig. 3 shows the mould assembly for one of these gearboxes and Fig. 4 the finished casting.

A gearbox housing for a helicopter propeller drive is depicted in Fig. 5. This casting is in alloy AZ 91, weighs approximately 300 kg and incorporates 8 meters of oil lines.

In car racing, as well as aerospace, weight savings and compact dimensions have top priority. Oil-sumps, camshaft housings, gearboxes and even water-cooled cylinder-blocks cast in magnesium have been successfully applied in Formula 1 and rally car engines. As an example, the oil-sump of a BMW Formula 1 racing engine is shown in Fig. 6.

To fulfill the high dimensional requirements of these castings, mould making techniques were greatly improved. All premium magnesium castings are noe made in cold-setting resin-bonded sand moulds. Furane or polyurethane resin binders are used for cores as well.

A basic change in pattern making was necessary to allow substantial reductions in casting tolerances. Hence wooden patterns have been replaced by metal or plastic patterns (if only a small number of castings is to be made from a pattern then mixed patterns are still used). By using core gauges, tolerances of wall thicknesses of ± 0.6mm for walls of 4-6mm can be achieved, and positional tolerances for passageways with branches in different joint faces of ± 0.8mm and even as narrow as ± 0.4mm are possible.

At the other end of the scale, magnesium castings are used for heavy and large components, e.g. castings for vibratory tables (Fig. 7).

Mg-Zr alloys were chosen for these castings because of their superior damping capacities. They are however very hard to cast owing to their small freezing range. A large size casting with minimum weight is shown with the aircraft frame in Fig. 8. Its dimensions are about 1500 x 1000 x 200 mm^3.

Permanent mould casting of magnesium alloys is usually applied in combination with low pressure casting units. Many car components are cast using this process, one example being wheels (Fig. 9). Even in the severe environment under which wheels are used corrosion can be successfully prevented by appropriate coatings.

In addition to the above mentioned processes magnesium is used for die casting and on a minor scale, for investment casting and low pressure sand casting.

FUTURE DEVELOPMENTS

In the second major section of this paper some aspects of the future development of magnesium casting will be discussed.

To increase the use of magnesium, further alloy development has to be carried out to meet the demands of the industry. The magnesium alloy producers have recently come up with some new alloys which still have to be fully applied.

Amongst the general purpose alloys AZ 91 HP has attracted increasing attention, because it shows markedly improved corrosion behaviour, in comparison with AZ 91 C, due to its low heavy metal (Fe, Ni and Cu) content. Fig. 10 compares corrosive weight losses of AZ 91 HP and AZ 91 C with those of an aluminium alloy. With this new alloy many of the corrosion problems which arise when applying magnesium alloys in a standard casting can be overcome.

For aerospace or special engineering purposes, newly developed high strength alloys like QE 22 or EQ 21 will have to replace thorium containing alloys like QH 21, because the use of radioactive materials will be further restricted in the future.
WE 54, an alloy which shows even more superior properties, will be used in cases where the highest requirements have to be met. Fig. 11 compares the elevated temperature properties of this alloy with those of high temperature aluminium alloys. The values at temperatures above 250°C are remarkable; exceeding those of high performance aluminium alloys. In addition to this WE 54 shows very good corrosion behaviour. To exploit its full potential on an industrial scale, further work needs to be carried out concerning its handling during melting and casting.

Fluxless melting of magnesium alloys is already applied by some foundries, but the results are not always satisfactory as the cleanliness of the cast metal from oxides and minor inclusions cannot yet be guaranteed under all circumstances. These problems are enhanced if returns are added to the charges. Improved filter systems need to be applied here because future disposal regulations will restrict the further use of fluxes.

The grain refining techniques of Mg-Al-Zn alloys will also have to be revised. There have been reports in Germany this year on the formation of HCB, where Hexachlorethane is used for the

degassing of aluminium melts. Similar chemical reactions are to be expected during the grain refinement of Mg-Al-Zn type alloys. One possible solution would be the reintroduction of the former practice of overheating, and subsequent cooling down, of the melts prior to casting. However, holding times necessary to carry out this process are uneconomic and cumbersome, therefore new grain refiners will have to be found.

To reduce wall thicknesses not only an increased number of gauges and more closely tolerated patterns will have to be used, but also the inhibitors added to the mould sands will have to be improved. Inhibitors will have to be found which further diminish the oxidation of the melt surfaces and mould-melt reactions, allowing an easier melt flow in the moulds and hence improving the filling capacity of the metal. These inhibitors ought to produce only the smallest amount of gas, minimizing blowholes or cold shuts.

Similar considerations apply to sand binders where it is necessary that a minimal amount of combustion gas is released, and when it is, as late as possible.

A casting method which is particularly suitable for thin walls, even for large areas, is low pressure casting. Here pressure - and not gravity - supports the metal flow in a controllable and reproducable fashion. Furthermore, feeding can be improved by an additional increase in pressure during solidification. This method will also lessen scrap returns because the extent of the running and gating systems, and risers, can be reduced. This process is currently applied on an industrial scale for aluminium with some sand and even more permanent-mould castings.

The principal arrangement of a low pressure casting unit is shown in Fig. 12. The metal is transported from the crucible into the mould by applying a gas pressure on the melt surface. The melt rises in the casting tube and fills the mould cavity without any turbulence, which is often created in a conventional running and gating system. The casting cycle is computer controlled and the casting unit can be designed to any degree of mechanisation.

Other designs of low pressure casting units incorporate electro-magnetic pumps (Fig. 13), which give an even better control of the metal flow. In particular the melt can be held much easier at a certain level in the casting tube than with a gas pressure system. This is very important with magnesium alloys as the movement of the metal level in the feed tube of such a casting unit ought to be limited, so reducing the oxidation of the melt. In addition to this, protective gas has to be applied to the melt surface in the feed tubes (this is not shown in the schematic drawings of Figs. 12 and 13).

Several companies are currently developing low pressure units for magnesium snad castings, but to our knowledge, low pressure casting of magnesium is at the moment only used commercially for permanent mould castings. The main problem with sand moulds is the interface between moulds and the casting tube, here special care (i.e. protective gas) has to be taken to avoid oxidation of the melt surface whenever moulds are removed after each casting cycle.

Another problem is the protection of the sand surface of the mould cavity. With increasing casting pressures, and in particular with increased added pressures for feeding, penetration of the melt into the sand mould might occur. This would be detrimental for the soundness of the casting surface. One solution might be to use the pressure system for filling purposes only.

Permanent mould casting of magnesium alloys will also benefit from the low pressure casting process. Several low pressure die casting units are already successfully used for the mass production of car and other components. These technologies will enable the foundries to produce high numbers of castings at competitive prices. This trend could be further supported by a more extensive use of magnesium pressure die casting.

Computerized simulation will have to be applied in the design of running and gating systems and moulds to make the development of castings more efficient. By simulating the solidification processes inside the mould cavity, the arrangement and size of chills in sand moulds, or the cooling and heating systems in permanent moulds, can be designed and therefore greatly reduce the number of trial castings. These procedures are already successfully used in aluminium casting and their application ought to be possible for magnesium casting as well.

Fig. 14 gives an example where the effect of different chill materials on the solidification of an A 357 alloy, in a critical area of an aerospace casting, is simulated.

CONCLUSIONS

The use of magnesium has been decreasing over the last two decades. This applies in particular to the automotive as well as the aerospace industries and the major reason for this is the lack of experience most users have with magnesium.

In spite of this unfavourable situation magnesium casting technology has been constantly improved. Compact castings for aerospace applications can be produced today which incorporate precast lubrication passageways of diameters as narrow as 2mm. These precast piping systems are able to stand pressures of up to 90 bar.

Wall thicknesses as thin as 3.5mm and casting tolerances of ± 0.5mm can be achieved.

If the foundry industry builds up on this present status and develops magnesium casting technology further, future demands can be met.

New alloys like AZ 91 HP and WE 54 will provide answers to the magnesium corrosion problem and the latter can offer improved mechanical properties at elevated temperatures, thus encouraging the further use of magnesium for automotive and aerospace applications.

Improved melt handling and mould making techniques will reduce the environmental hazards of magnesium casting, allowing the production of sounder castings to closer tolerances, which in turn, will be enhanced by the application of new casting processes, e.g. low pressure sand casting. The latter would permit the production of castings with even thinner walls than are possible today, in a more controllable fashion, and reduce scraps and returns thus increasing economy.

Low pressure die casting and pressure die casting can help to increase mass production of magnesium castings and thus boost the use of magnesium.

A further important contribution towards a more cost efficient production will come from the application of computerized mould design and casting layout. Simulation of the solidification of castings and thermal processes in the moulds will speed up casting development.

If the above steps are taken, magnesium casting will become more competitive which will result in an increased use of magnesium and its alloys. The future of magnesium will therefore be brighter.

REFERENCES

1) I. J. Polmaer,
   Light Alloys, Metallurgy of the Light Metals,
   Edward Arnold Ltd., London, 1981

2) Norsk Hydro,
   Pamphlet on Pressure Die Casting of
   Magnesium Alloys

3) H. J. Dundas,
   AMAX Publication, AMAX Mg Div.,
   Salt Lake City, UT 84116, USA

4) W. Unsworth and J. King,
   New High Performance Casting Alloy Developed,
   Metallurgia, May 1986

Rm
[N/mm²]

* values according to AMS
** average cutup values

500
400
300
200
100

Rp 0,2
[N/mm²]

500
400
300
200
100

A (%)

8
6
4
2

A 201 - T7 *
A 206 - T4 *
A 357 - T6 **
QE 22A - T6 **
ZE 41A - T5 **
AZ 91 - T6 **

Al - Alloys   Mg - Alloys

1   Comparison of mechanical properties of
    aluminium alloys with magnesium alloys

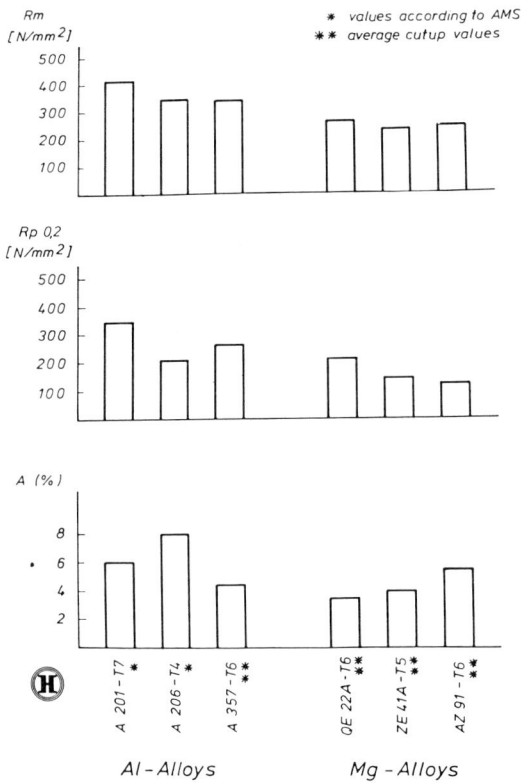

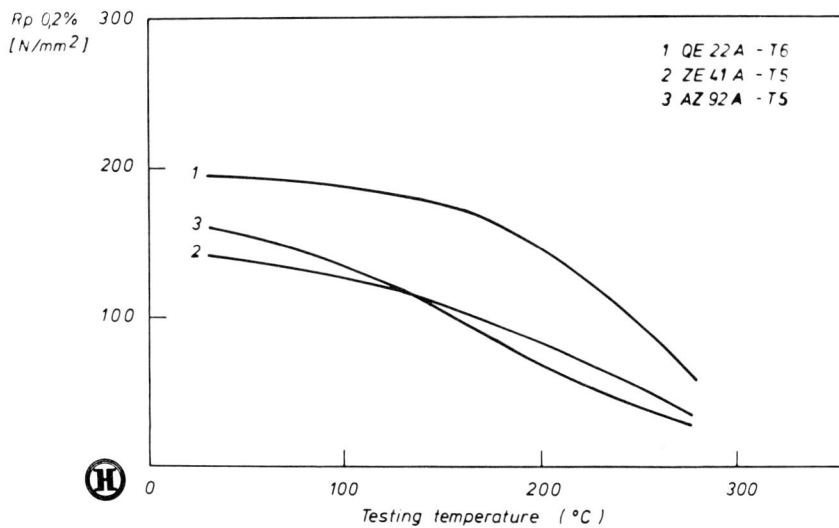

3   Mould assembly for gearbox housing of an
    auxiliary power unit for aeroplanes

Rp 0,2%   300
[N/mm²]

1 QE 22A - T6
2 ZE 41A - T5
3 AZ 92A - T5

200

1
3
2

100

0   100   200   300

Testing temperature (°C)

2   Yield stresses of magnesium alloys at
    elevated temperatures (1)

4 Finished casting of gearbox housing for auxiliary power unit in Mg-Zn-RE-Zr type alloy (ZE 41 S)

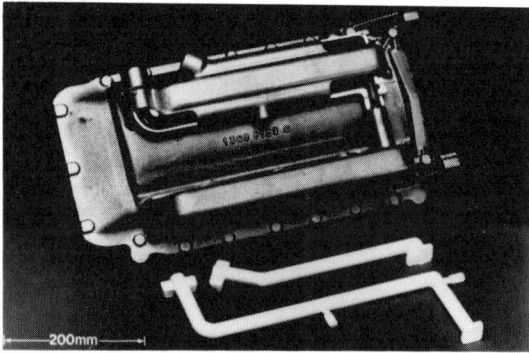

6 Oil-sump of a BMW Formula 1 racing engine (courtesy of BMW Motor Works)

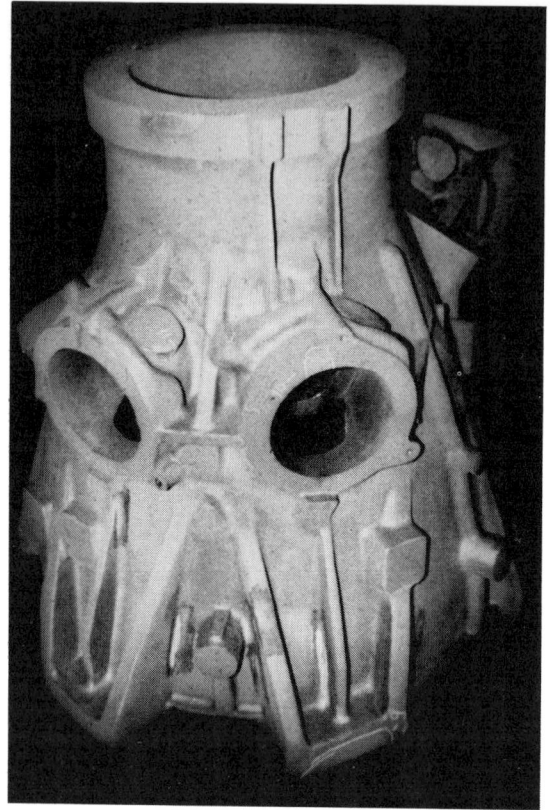

5 Gearbox housing for a helicopter propeller drive (courtesy of Fansteel, Wellmann Dynamics)

7 Castings for a vibratory table (adapter 120 kg, table 340 kg)

8 Aircraft frame (size 1500 x 1000 x 200 mm^3)

9   Car wheels made out of magnesium alloys (2)

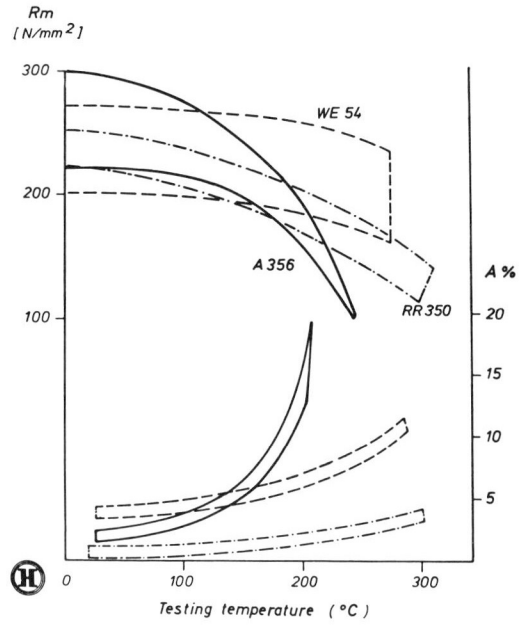

11   High temperature properties of WE 54 compared
     with high strength aluminium alloys (4)

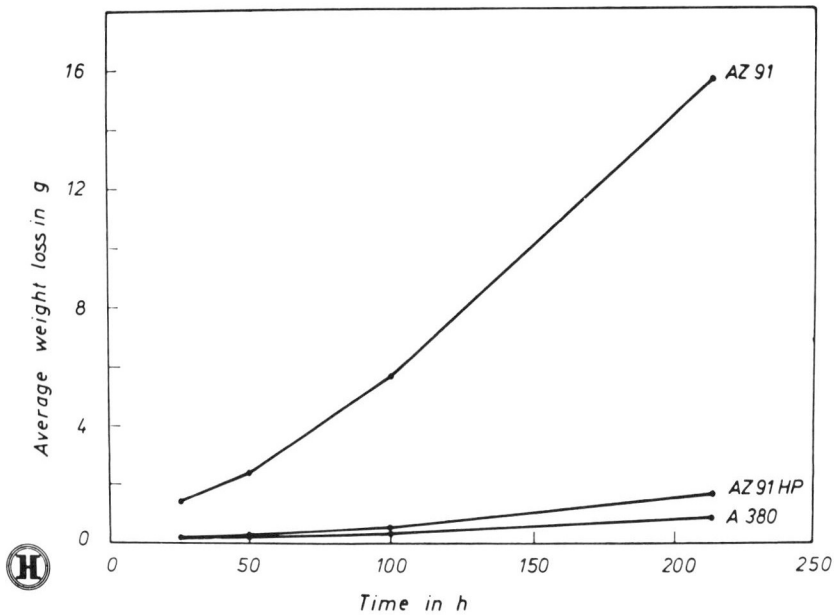

10   Comparison of the corrosion behaviour of
     AZ 91 C and AZ 91 HP with the aluminium
     alloy A 380 (3)

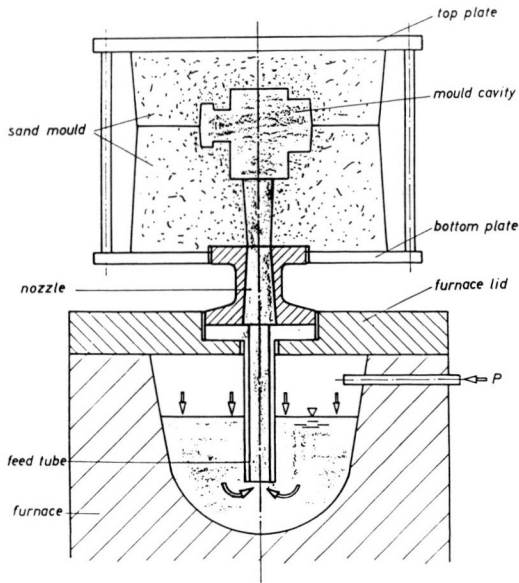

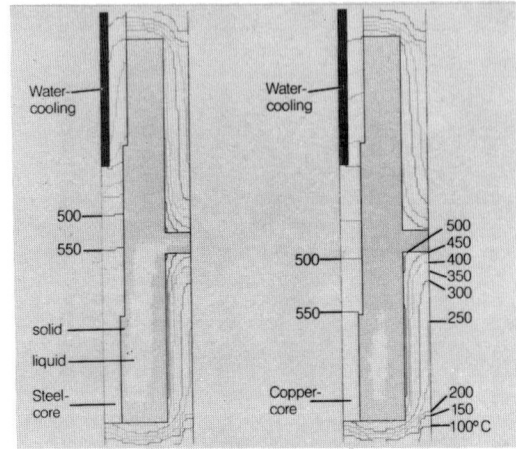

14 Comparative simulation of the solidification of an aluminium alloy, in a critical area of an aerospace casting, using different chill materials

12 Schematic arrangement of a gas pressure operated low pressure sand casting unit

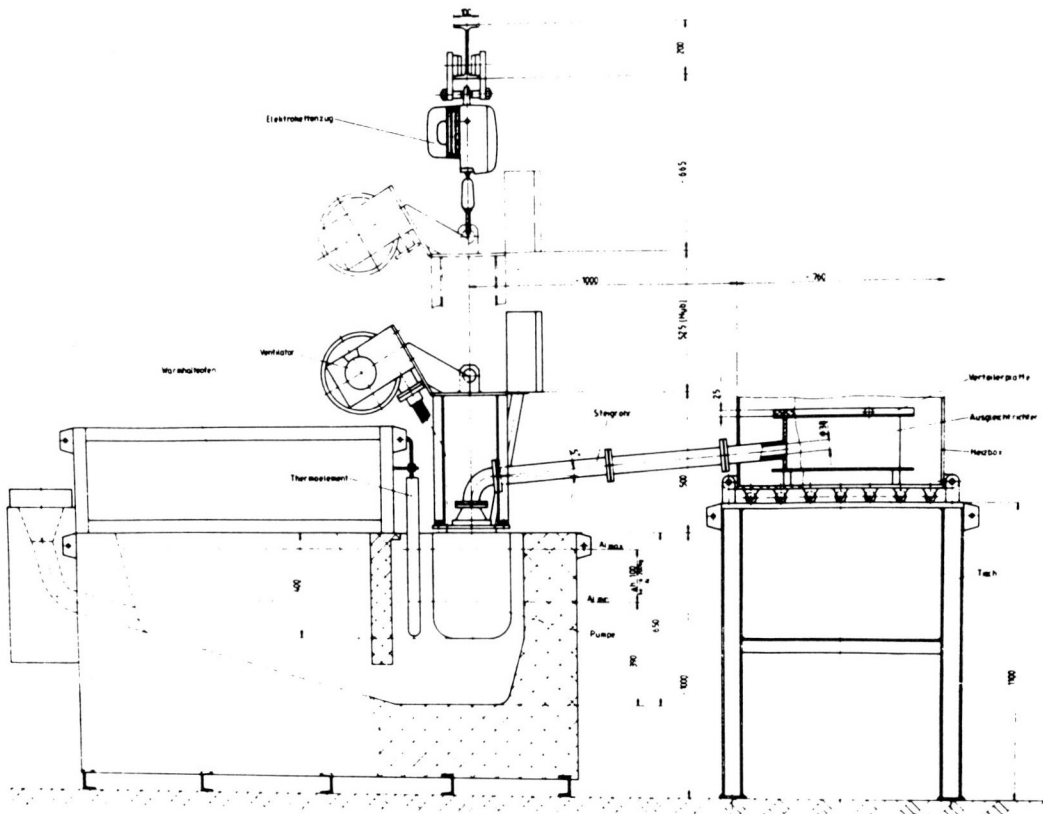

13 Aluminium low pressure sand casting unit (schematic) with electromagnetic pump

Recent casting alloy developments

W UNSWORTH and J F KING

The authors are both with Magnesium Elektron Limited, Manchester, England.

SYNOPSIS

Alloy systems Mg-Zn-Cu-(Mn) and Mg-Y-Nd-(Zr) have been recently investigated. The effect of alloying additions and heat treatment on the structure, precipitation processes and mechanical properties of alloys in both systems are described.

A practical casting alloy in the Mg-Zn-Cu-Mn system has been developed with good castability and moderate strength at temperatures up to at least 150oC. In the Mg-Y-Nd-Zr system a high strength alloy with excellent temperature stability up to 300oC has been developed with corrosion resistance comparable to that of high strength and creep resistance aluminium based casting alloy.

INTRODUCTION

Magnesium alloys in current use can be divided into two major groups; those based on the Mg-Al system and those refined with zirconium.

The Mg-Al series of alloys provides satisfactory ambient temperature properties but elevated temperature capability is limited.[1] While the creep resistance of Mg-6 wt % Al-1 wt % Zn (AZ61)[2] and Mg-9 wt % Al-2 wt % Zn (AZ92)[3] can be improved by calcium and silicon additions respectively, the improvement obtained is modest. There appear to be limited options for further improvement in properties by alloying additions.

A greater number of elements can be used in zirconium refined alloys to achieve a wide range of properties.[4] Moderate strength and excellent temperature resistance can be obtained in alloys containing zinc as the hardening addition in conjunction with cerium rich rare earth metals (mischmetal) or thorium while higher ambient temperature properties and moderate temperature resistance are achieved in alloys containing silver with the more soluble neodymium.

The combination of high ambient temperature and elevated temperature resistance is not achievable with any of these alloys.

Because of additional processing requirements with zirconium refined alloys, these are more expensive than the Mg-Al type alloys and their use tends to be restricted to aerospace/military or premium commercial applications. The Mg-Al alloys are widely used for general commercial applications but here a requirement exists for an alloy with better temperature stability than that offered by the Mg-Al alloys.

Two alloy systems were investigated with the objective of meeting these requirements.

MAGNESIUM-ZINC-COPPER SYSTEM

Background

While zinc is an effective solid solution hardener, the optimum zinc content in the as-cast binary alloy is around 6 wt % but in the fully heat treated alloy this is reduced to 4 wt %.[5] Binary alloys are brittle and hot short but ductility can be improved by zirconium refining which also raises the optimum zinc content in the fully heat treated alloy to between 5 and 6%.[6] The ternary alloys are however prone to layer microshrinkage and are not weldable which limits their usefulness. Addition of rare earth metals or thorium to the ternary alloy eliminates microshrinkage and confers weldability on the alloys; this is the basis of several currently used alloys.

The more complex alloys containing rare earth metals and zirconium are more expensive than those based on the Mg-Al-Zn-(Mn) system such as AZ91 (Mg-9 wt % Al-1 wt % Zn) and another method of overcoming the limitations of high zinc alloys has been investigated by U.S. researchers. Using aluminium as a major alloying addition to high zinc binaries, compositions offering good ambient and elevated temperature properties in the pressure die cast condition were developed.[7] However when sand cast the properties obtained from these alloys are low and inferior to those obtainable with AZ91.

Copper additions to binary Mg-Zn alloys were found to markedly improve ductility and give alloys having a good solution heat treatment response.

Effect of copper on structure and heat treatment response of Mg-Zn alloys

Progressive additions of copper to Mg-Zn binary alloys significantly increase the eutectic temperature of the alloy but the effect becomes less marked with the higher, e.g. 8-10 wt %, Zn levels (Fig. 1). This increase in eutectic temperature permits solution treatment temperatures to be increased as the copper content is increased thereby ensuring maximum solution of zinc and copper.

Alloys containing above about 1 wt % copper are relatively fine grained, the grain sizes in the sand-cast condition comparing favourably with those of fully grain refined Mg-Al-Zn alloys. The grain refining effect does not deteriorate on repeated remelting.

In binary Mg-Zn alloys the eutectic is completely divorced with Mg-Zn compound around the grain boundaries and between dendrite arms. With the addition of copper the eutectic becomes truly lamellar (Fig. 2a); the intermetallic phase having a diamond cubic structure of composition $Mg(Cu.Zn)_2$. [8] On solution treatment partial dissolution of the eutectic occurs leaving rounded rods or platelets within an α matrix, (Fig. 2b). Such a structure would be expected to improve ductility of the alloy.

As the zinc and copper content increases, the amount of eutectic phase in the as-cast condition increases suggesting improved castability with increasing alloy content.

Optimum ageing response is obtained with alloys containing around 1.5 wt % copper; at higher levels the ageing response is impaired.

Two main precipitating species, needle and plate forms, have been identified in the Mg-Zn-Cu alloys and these appear identical with those in binary Mg-Zn alloys.[9] The precipitating sequence is shown in Table 1.

Peak hardening occurs at around 200°C and both the β_1 and β_2 precipitates contribute to the hardening effect; the addition of manganese stabilises the ageing response. In the absence of manganese, alloys overage rapidly at 250°C; the presence of manganese slows down the speed of overageing.

Effect of Composition on mechanical properties of Mg-Zn-Cu alloys

In the as-cast condition the yield strength is low, 55-85 N/mm2; the levels achieved being more dependent on copper than zinc content. Copper additions also improve tensile strength and for alloys containing 0.5-3.5 wt % Cu this ranges from 165-200 N/mm2 depending on zinc content. Elongation values increase progressively with increasing copper content.

Solution heat treatment at a temperature 25-30°C below the solidus temperature followed by hot water quenching and precipitation heat treatment at 180°C results in a large increase in yield strength, more than doubling the as-cast value. Tensile strength is also increased and elongation decreases as would be expected but the decrease in ductility is small in relation to the magnitude of the improvement in yield strength.

In the fully heat treated condition mechanical properties are highly dependent on levels of zinc and copper (Fig. 3). Yield strength increases with increasing zinc and copper contents attaining maximum values at around 9 wt % zinc; maximum tensile strength occurs at somewhat lower zinc levels. The optimum copper content for maximum strength is 1.5-2.0 wt % which is in accord with the results obtained from structural investigations. Ductility, which improves with increasing copper content at all levels investigated, attains its maximum value at intermediate zinc levels (approx. 6 wt %) and copper contents about 2.5 wt %.

Effect of fourth element additions on mechanical properties of Mg-Zn-Cu alloys

The effect of fourth element additions to Mg-Zn-Cu alloys was examined with a view to further improving mechanical properties. Manganese, Aluminium and Silicon were the only elements which had a significant effect on fully heat treated alloys.

Manganese additions result in a significant improvement in yield strength in the fully heat treated condition and minimise the fall-off in strength above the optimum 1.5 wt % Cu (Fig. 4). Some loss in tensile strength occurs particularly with lower copper contents and ductility is reduced at all copper levels.

Aluminium, even at levels as low as 0.5 wt %, significantly reduces the eutectic temperature of the alloys and this prohibits the application of a high temperature solution treatment necessary to generate optimum properties.

Silicon additions of as little as 0.2 wt % significantly reduce yield strength.

Development of Practical Casting Alloys

Results from mechanical and structural property investigations indicated two possible alloy options:-

(1) for maximum strength the alloys should contain at least 8.5 wt % Zn and at least 1.5 wt % Cu.

(2) for maximum ductility the zinc should be reduced to around 6 wt % and the copper content increased to around 2.5 wt %.

Since the end product is a cast shape, castability is an important consideration, particularly where complex or thin wall components are being produced. Comparative trials with compositions around the indicated options showed that the lower zinc-higher copper alloys gave better fluidity and freedom from hot cracking and microshrinkage compared with the high strength versions. To offset the inevitable loss of strength from lowering of the zinc content manganese was added to give a compositional range:

Zinc	5.5	- 6.5	wt %
Copper	2.4	- 3.0	wt %
Manganese	0.25	- 0.75	wt %

Properties of Mg-6 Zn - 2.5 Cu-(Mn) [ZC63] alloy

Properties obtained on separately cast test bars (sand cast) are summarised in Tables 2-5 inclusive.

Overall the properties are better than those obtainable in currently used Mg-Al-Zn alloys but are not as good as those from the zirconium grain refined alloys. Retention of strength at elevated temperatures is better than that of the Mg-Al-Zn alloys and this is also reflected in better creep strength.

Fatigue strength in the un-notched condition is better than that of Mg-Al-Zn alloys in the fully heat treated condition while the notched values are comparable.

Copper additions to Mg-Al-Zn alloys have a detrimental effect on corrosion resistance.[10] This is not the case with Mg-Zn-Cu alloys and tests under salt immersion conditions have shown the corrosion resistance of the ZC63 alloy to be better than that of Mg-Al-Zn alloy AZ91-C in the fully heat treated condition, i.e 4.7 mgms/cm2/day compared with 7.3 mgms/cm2/day for AZ91C.

A number of castings have been produced under practical foundry conditions using sand, gravity die and precision casting techniques. These have confirmed the good casting characteristics of the alloy, its freedom from microshrinkage and its ability to be welded using the tungsten arc inert gas technique.

Properties obtained on specimens cut from castings confirm the levels obtained on separately cast test bars. Furthermore properties in casting show more uniformity and appreciably less variation with section thickness than those made in Mg-Al-Zn alloys.

MAGNESIUM-YTTRIUM-NEODYMIUM SYSTEM

Background

Addition of zirconium to binary magnesium-cerium rich rare earths (mischmetal) alloys improves elevated temperature characteristics.[11] Substitution of cerium rich rare earths by more soluble neodymium produces alloys which respond to full heat treatment and combine good ambient and elevated temperature properties.[12] Castability of the ternary alloys is poor; addition of zinc to the mischmetal containing alloy and silver to the neodymium containing alloy gives improved mechanical properties and better castability.

Yttrium has a high solubility in magnesium (12.5 wt %) and the progressive decrease in solubility with decreasing temperature should permit an age hardening response. Alloys based on Mg-Y-Zn-Zr[13] and Mg-Y-Nd-Zn-Zr[14] appear in patent literature.

These claim good elevated temperature properties but ambient temperature strength is moderate and the alloys do not respond to a full solution precipitation heat treatment. This is probably because a major proportion of the yttrium remains tied up as a stable Y-Zn compound preventing solution/precipitation reactions. However addition of yttrium to Mg-Nd-Ag-Zr alloys gave

fully heat treatable alloys with both high ambient and elevated temperature strength although at least 4 wt % yttrium was necessary to obtain the full effect.[15]

It therefore appeared that fully heat treatable alloys combining high strength with good temperature resistance could be developed in the Mg-Y-Nd system.

Effect of Nd and Y on structure and heat treatment response

Additions of Yttrium and neodymium result in a progressive increase in eutectic content of the as-cast alloys.

Yttrium has an appreciable grain refining effect but this is less marked than zirconium's. In the presence of zirconium the eutectic has a more irregular appearance and is also lameller in contrast to other grain refined alloys (Fig. 5a).

On solution treatment the eutectic is almost completely dissolved (Fig. 5b); the amount remaining depends more on the neodymium content than the yttrium content as would be expected.

On subsequent precipitation treatment only one hardness peak has been identified.[16] This develops slowly at low temperatures (175°C), attains peak hardness at around 200°C and is barely discernable at 350°C. The precipitation sequence is shown in Table 6.

The β_1'' precipitate forms at the lower temperature (approx. 175°C) with an ordered hexagonal structure of the DO_{19} (Mg$_3$Cd) type and slowly transforms at the higher temperature (200°C) to β_1'. The latter has a base centre orthothombic structure. There is no change in the structure at 250°C but the precipitates are resolved as platelets. At temperatures above 300°C the β_1' converts to the β form having a face centre cubic structure. This is tentatively assumed to be the equilibrium precipitate and appears to be a ternary compound.

Composition of the intermediate precipitates, which are responsible for the good ageing response, has not been established.

Effect of composition on properties of Mg-Y-Nd-Zr alloys

The effect of varying the yttrium and neodymium content on the yield and tensile strength of the fully heat treated quarternary alloys is shown in Fig 6. Strength increases with increasing yttrium and neodymium content while ductility decreases, neodymium content having the greater effect on properties.

To achieve maximum strength with acceptable ductility the alloy should contain around 6 wt % of yttrium and around 2 wt % neodymium.

Development of a Practicable Casting Alloy

Yttrium is expensive, particularly when used as a major alloying addition and the use of cheaper, less pure, grades is commercially desirable.

Yttrium is widely distributed in association with other rare earth metals in ores such as xenotime, monazite etc. and yttrium rich concentrates of varying content can be obtained as intermediate products during the processing of these ores.

An 'yttrium mischmetal' derived from concentrates containing 75 wt % yttrium with the balance heavy rare earths, mainly Dysprosium, Erbium, Ytterbium and Gadoliniun, has proved to be an effective substitute for pure yttrium.

Substitution does not affect the precipitation sequence and results in only a slight reduction in strength. The heavy rare earths added with the yttrium appear to restrict the solubility of neodymium and some reduction in the latter addition is necessary to achieve a satisfactory compromise of tensile strength and castability, i.e.

Yttrium   5 - 5.5 wt % (plus HRE elements)
Neodymium 1.5 - 2 wt %
Zirconium 0.3 - 0.6 wt %

The conventional method of melting magnesium alloys relies on the use of fluxes, based on alkali and alkaline earth halides, to protect against oxidation.

Such techniques cannot be used on alloys containing appreciable amounts of yttrium since this reacts with $MgCl_2$, an essential constituent of these fluxes, resulting in unacceptable loss of yttrium. High yttrium losses are also encountered with fluxless techniques using CO_2/SF_6 gas mixtures now being used successfully on other magnesium base alloys. Alloys are therefore processed in an inert atmosphere (Argon + 0.5 vol % SF_6) although standard CO_2/SF_6 protection techniques can still be used at the casting stage. A processing technique using this gas mixture in simple closed crucibles has been proven under practical foundry conditions (Fig. 7).

Properties of Mg-5Y-3.5 TRE-Zr (WE54) alloy

Ambient temperature properties from separately cast test bars and specimens cut from castings are very similar (Table 7) and these properties are little affected by short term exposure at temperatures up to 300ºC. In this respect the alloy is superior to currently used high strength Mg-Ag-Nd-Zr alloy (QE22) and compares favourably with some of the high strength aluminium casting alloys (Fig. 8).

On long term temperature exposure (10,000 hours at 250ºC) the alloy is clearly superior to any currently used magnesium alloys (Fig. 9) and again compares favourably with RR350.

The alloy is more creep resistant at temperatures up to 250ºC than the currently used high strength magnesium alloys and similar to that of the thorium containing alloys HK31 (Mg-3Th-Zr) and HZ32 (Mg-3Th-2.5 Zn-Zr) (Fig. 10). Stress to rupture in 1000 seconds at 250ºC is 210 N/mm^2 which is almost double that of QE22.

Fatigue endurance limit on rotating/bending tests at 5 x 10^7 cycles is 97 N/mm^2 and of the same order as that of QE22 and A356. At 250ºC the fatigue endurance value is 75 N/mm^2 which is significantly better than QE22 (50 N/mm^2).

Corrosion resistance is markedly superior to that of the other high temperature magnesium alloys,

approaching that of some currently used aluminium casting alloys (Table 8). No evidence of stress corrosion cracking has been observed with stress levels up to 100% of the yield strength.

A range of castings have been made under practical foundry conditions using sand, gravity die and precision casting techniques. These have confirmed the good casting characteristics of the alloy and its ability to be welded using tungsten arc inert gas techniques.

CONCLUSIONS

Two new alloy systems have been investigated and practical casting alloy compositions have been developed in each system.

ZC63 combines moderate ambient temperature strength with useful elevated temperature properties up to at least 150ºC. The alloy has good castability, is weldable, and free from layer microshrinkage, enabling pressure tight castings to be produced.

The alloy should be of interest for high temperature automotive applications e.g. engine components, impellers etc.

WE54, combining high ambient temperature strength, excellent thermal stability and good corrosion resistance is particularly suited to aerospace military applications, e.g. gear boxes designed to operate at high temperature.

Some limited improvement in properties by compositional and heat treatment modifications may be possible in both alloy systems but it is believed that these will be relatively small. Opportunities for further alloy development appear greatest for alloys aimed at high temperature applications where some of the rare earth elements, currently available in commercial quantities, may be capable, given suitable heat treatment, of producing temperature stable precipitates.

REFERENCES

(1) W. Unsworth, J.F. King. Proc. AGARD Conf. No. 325 4-9th April 1982. (AGARD-CP-325) Paper 5.

(2) I.I. Filippov Liteznoje Proiswodstwo 1960, 2.5.6 10-13.

(3) British Patent 847992.

(4) W. Unsworth, Light Metal Age 1986 44 7.9.15.

(5) F.A. Fox, J. Inst. Met. 1945 71 415.

(6) J.W. Meir, Trans A.F.S. 1952 60 603.

(7) G.S. Foerster, Proc. 33rd Int. Magnesium Association Meeting, Montreal 23-25 May 1976 p 35.

(8) R.I. Moss, PhD Thesis University of Manchester 1983.

(9) J.S. Chunn, J.G. Byrne, J. Mat. Sci 1969 4 1981.

(10) J.E. Hillis, Light Metal Age June 1983 25.

(11) A.J. Murphy, R.J.M. Payne, J. Inst. Met. 1947 73 105.

(12) K.E. Nelson, Trans A.F.S. 1959 67 601.

(13) G.S. Foerster, J.B. Clark, U.S. Patent 3,419,385 1968.

(14) N.M. Tikhova et al. B. Patent 1,378,281 1973.

(15) W. Unsworth, J.F. King, S.L. Bradshaw, British Patent 1,463,609 1976.

(16) H. Karimzadeh, PhD Thesis, University of Manchester 1985.

TABLE 1. PRECIPITATION SEQUENCE IN Mg–Zn–Cu ALLOYS

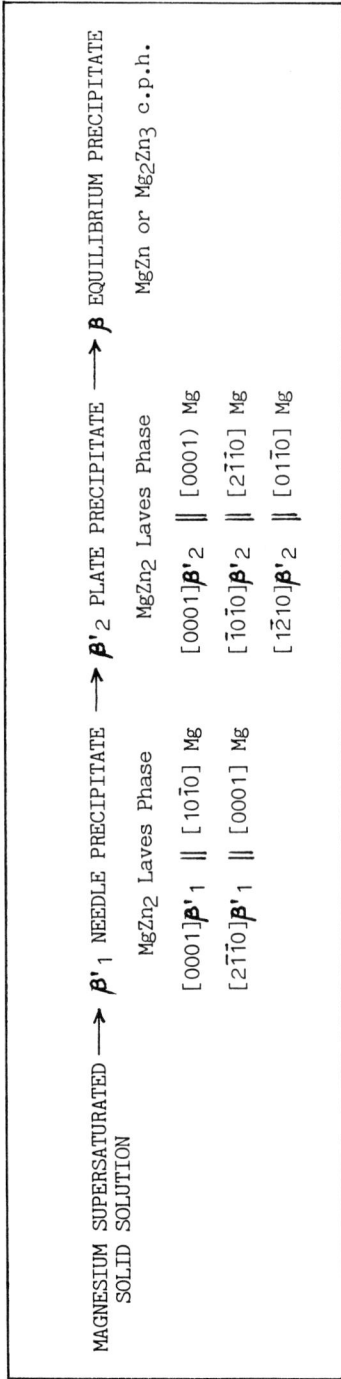

MAGNESIUM SUPERSATURATED $\longrightarrow$ β'_1 NEEDLE PRECIPITATE $\longrightarrow$ β'_2 PLATE PRECIPITATE $\longrightarrow$ β EQUILIBRIUM PRECIPITATE
SOLID SOLUTION

MgZn$_2$ Laves Phase	MgZn$_2$ Laves Phase	MgZn or Mg$_2$Zn$_3$ c.p.h.
$[0001]\beta'_1 \parallel [10\bar{1}0]$ Mg	$[0001]\beta'_2 \parallel [0001]$ Mg	
$[2\bar{1}\bar{1}0]\beta'_1 \parallel [0001]$ Mg	$[\bar{1}0\bar{1}0]\beta'_2 \parallel [2\bar{1}\bar{1}0]$ Mg	
	$[1\bar{2}10]\beta'_2 \parallel [01\bar{1}0]$ Mg	

TABLE 2 – AMBIENT TEMPERATURE TENSILE PROPERTIES OF ZC63

Separately Cast Test Bars – Fully Heat Treated*

0.2% Yield Stress	130–160 Nmm^{-2} (19–23 K.S.I.)
Tensile Strength	215–260 Nmm^{-2} (31–38 K.S.I.)
Elongation	3–8%

*Heat Treatment: 8 hours at 440°C
Hot Water Quench
16–24 hours at 180–200°C

TABLE 3 – ELEVATED TEMPERATURE TENSILE PROPERTIES OF ZC63

Typical Tensile Properties from Separately Cast Test Bars
(fully heat treated)

Temperature °C	0.2% Proof Stress Nmm^{-2} (KSI)	Tensile Strength Nmm^{-2} (KSI)	Elongation %
20	158 (22.9)	242 (35.1)	4.5
100	141 (20.4)	215 (31.2)	9
150	134 (19.4)	179 (26.0)	14
200	118 (17.1)	142 (20.6)	11

TABLE 4 – CREEP RESISTANCE OF ZC63 AT 150 AND 200°C

Constant load test on specimens machined from separately cast test bars
(fully heat treated)

Test Temp °C	Time (Hours)	Stress to Specified Total Strain – Nmm⁻² (KSI)			
		0.1%	0.2%	0.5%	1.0%
150	10	42 (6.1)	77 (11.2)	101 (14.6)	
	100	37 (5.4)	73 (10.6)	100 (14.5)	
	1000	27 (3.9)	61 (8.8)	91 (13.2)	
200	10		62 (9.0)	ND	ND
	100		55 (8.0)	65 (9.4)	67 (9.7)
	1000		41 (5.9)	53 (7.7)	57 (8.3)

TABLE 5 – FATIGUE STRENGTH OF ZC63

Rotating bending tests on specimens machined from separately cast test bars
(fully heat treated)

ENDURANCE LIMIT Nmm⁻² (KSI)	STRESS REVERSALS			
	10^6	5×10^6	10^7	5×10^7
UN-NOTCHED	100 (14.5)	94 (13.6)	92 (13.3)	90 (13.1)
U –NOTCHED (SCF 2)	62 (9.0)	57 (8.3)	56 (8.1)	55 (8.0)

TABLE 6 – PRECIPITATION SEQUENCE IN Mg-Y-Nd-Zr ALLOYS

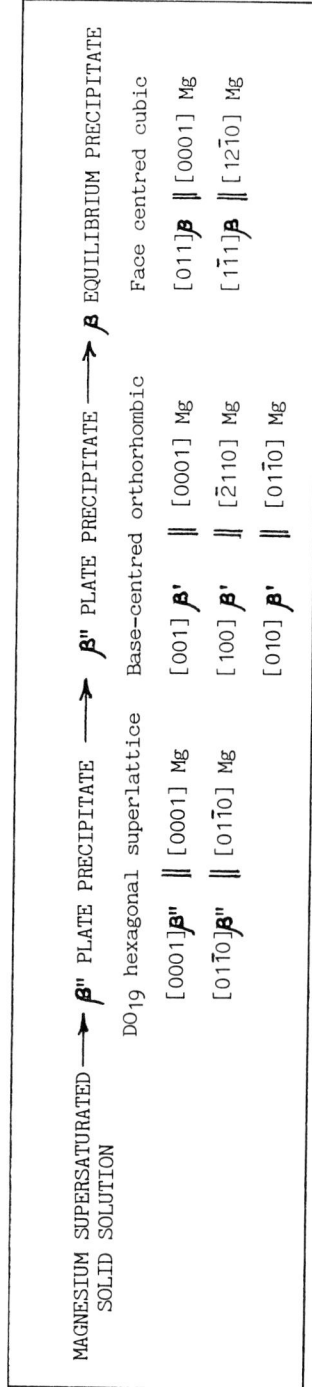

MAGNESIUM SUPERSATURATED SOLID SOLUTION ⟶ β″ PLATE PRECIPITATE ⟶ β″ PLATE PRECIPITATE ⟶ β′ PLATE PRECIPITATE ⟶ β EQUILIBRIUM PRECIPITATE

DO₁₉ hexagonal superlattice	Base-centred orthorhombic	Face centred cubic
$[0001]\beta'' \parallel [0001]$ Mg	$[001]\beta' \parallel [0001]$ Mg	$[011]\beta \parallel [0001]$ Mg
$[01\bar{1}0]\beta'' \parallel [01\bar{1}0]$ Mg	$[100]\beta' \parallel [\bar{2}110]$ Mg	$[\bar{1}\bar{1}1]\beta \parallel [1\bar{2}10]$ Mg
	$[010]\beta' \parallel [01\bar{1}0]$ Mg	

31

TABLE 7 – ANALYSIS OF TENSILE DATA FOR WE54-T6 TEST BARS AND CASTINGS

MELTS	SPECIMENS	VALUE	TENSILE PROPERTIES		
			YIELD STRESS Nmm^{-2} (KSI)	U.T.S. Nmm^{-2} (KSI)	% ELONGATION
SPECIMENS CUT FROM SEPARATELY CAST BARS					
94	261	max min mean	225 (32.6) 176 (25.5) 200 (29.0)	315 (45.7) 242 (35.1) 279 (40.5)	11 1 4
SPECIMENS CUT FROM CASTNGS*					
22	246	max min mean	254 (36.8) 168 (24.4) 207 (30.0)	304 (44.1) 214 (31.0) 273 (39.6)	11 0.5 4

*Specimens taken from random locations; not "specified" areas

TABLE 8 – COMPARATIVE CORROSION RESISTANCE OF WE54-T6 AND OTHER MAGNESIUM AND ALUMINIUM BASE CASTING ALLOYS

ALLOY TYPE/DESIGNATION		SEAWATER IMMERSION (28 DAYS) mgm/cm^2/day weight loss	ASTM B-117 SALT FOG (10 DAYS) mils/year penetrations
Magnesium Casting Alloys	WE54-T6	0.08 – 0.2	5 – 15
	ZE41-T5; EZ33-T5	2 – 4	350 – 550
	AZ91C-T6	6-10	600 – 1200
Aluminium Casting Alloys	C355, A356, A357	0.04 – 0.08	2 – 8
	A201, A206, RR350	–	10 – 20

32

Effect of Cu in Mg-Zn-Cu Alloys on Solidus Temperature

FIGURE 1

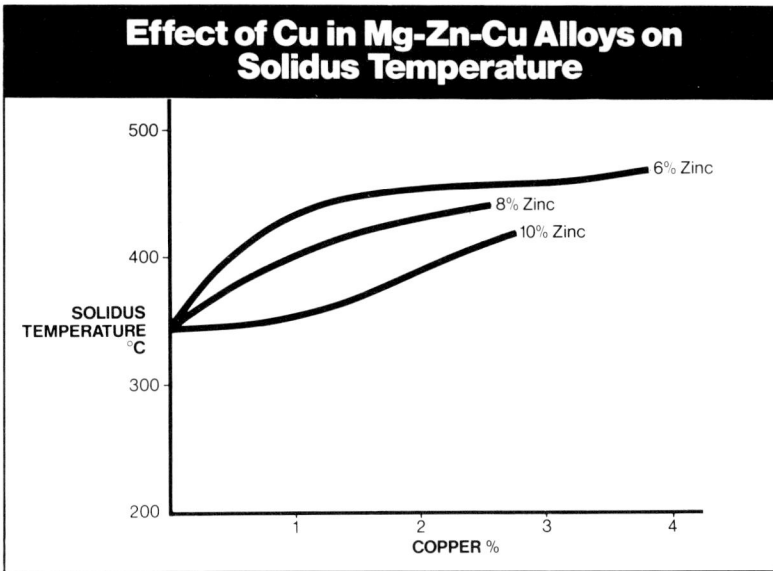

FIGURE 2

Microstructure of Mg-6% Zn-3% Cu Alloy
(original magnification X400)

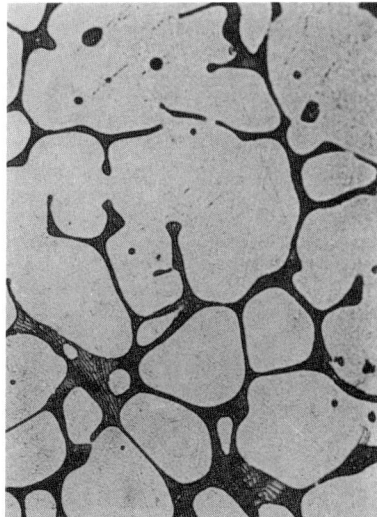

(a) As-cast

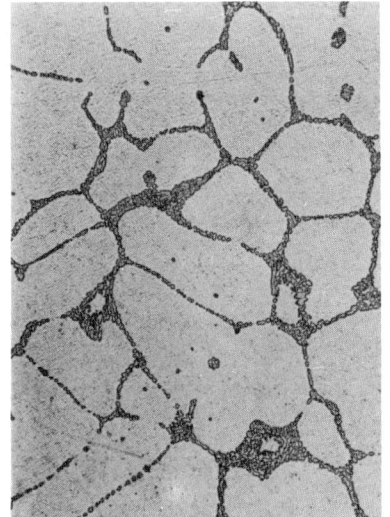

(b) After full heat treatment

Effect of Composition on Yield and Tensile Strength of Mg-Zn-Cu Alloys

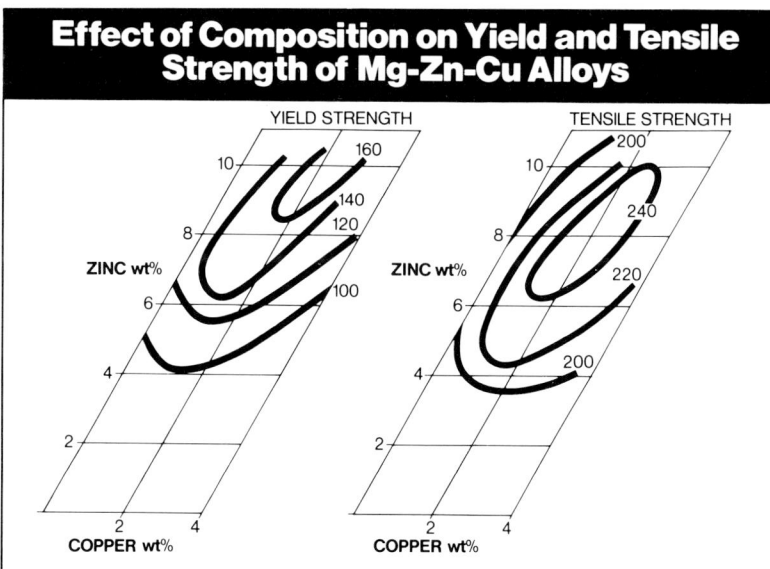

FIGURE 3

Effect of Copper Content on Tensile Properties of Mg-6% Zn Alloy Fully Heat Treated

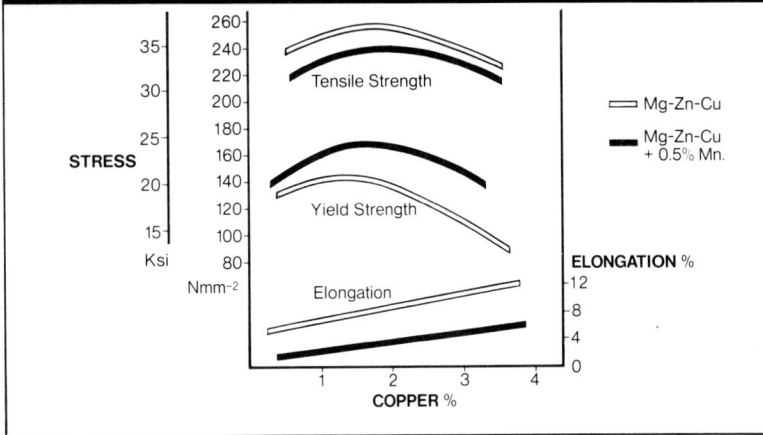

FIGURE 4

FIGURE 5

Microstructure of Mg-Y-RE-Zr Alloy(WE54)

(original magnification X200)

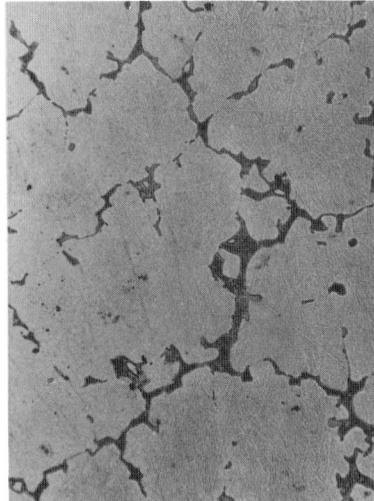

(a) As-cast

(b) After full heat treatment

Effect of Composition on Yield and Tensile Strength of Mg-Y-Nd-Zr Alloys

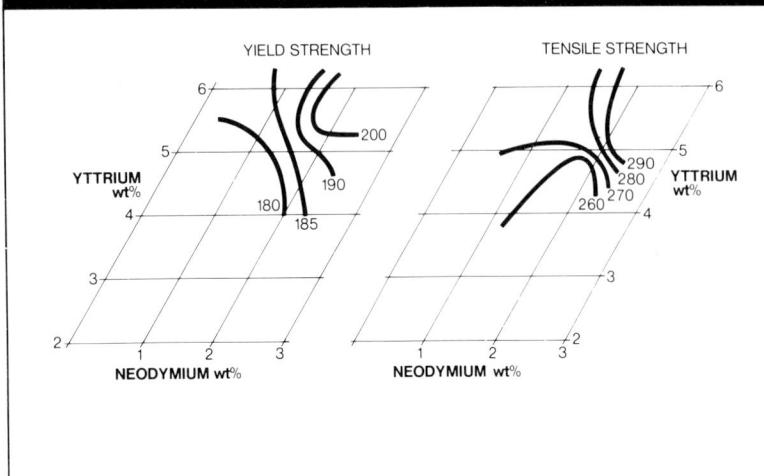

FIGURE 6

Experimental Scale Melting/Casting Unit

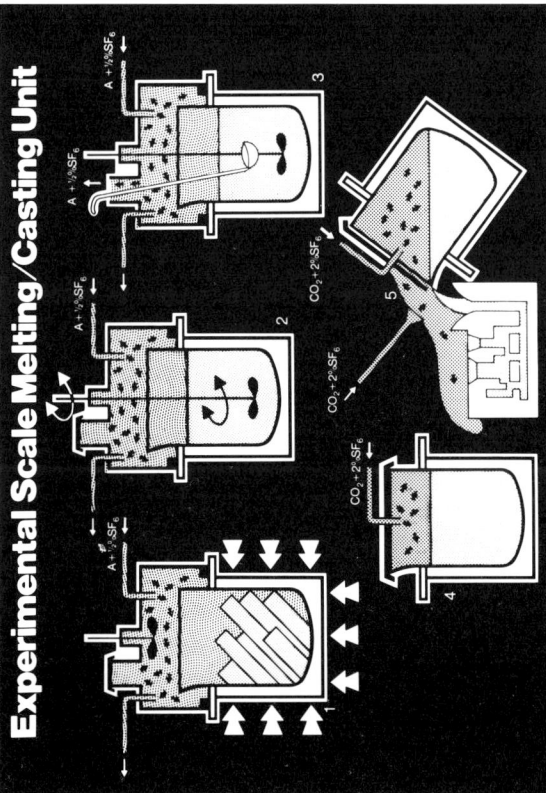

FIGURE 7

Effect of Temperature on Yield Strength of WE54 V A356, RR350 and QE22

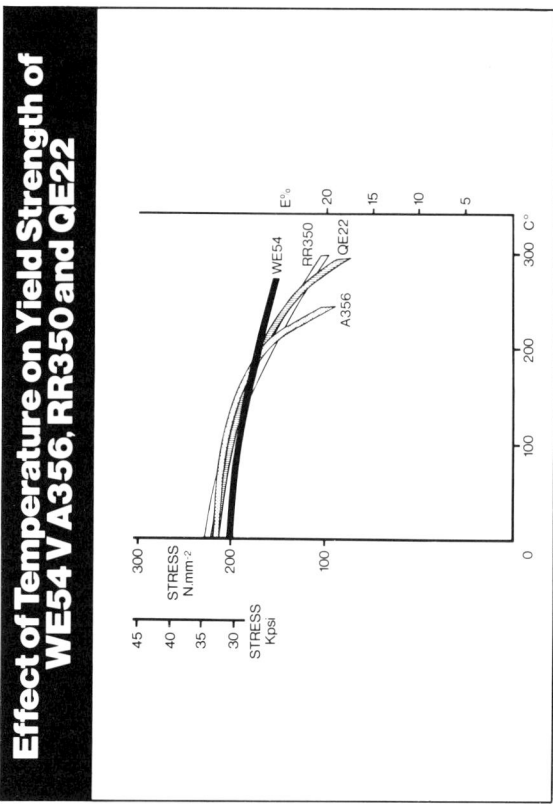

FIGURE 8

Effect of Exposure To Temperature on 0.2% Proof Stress

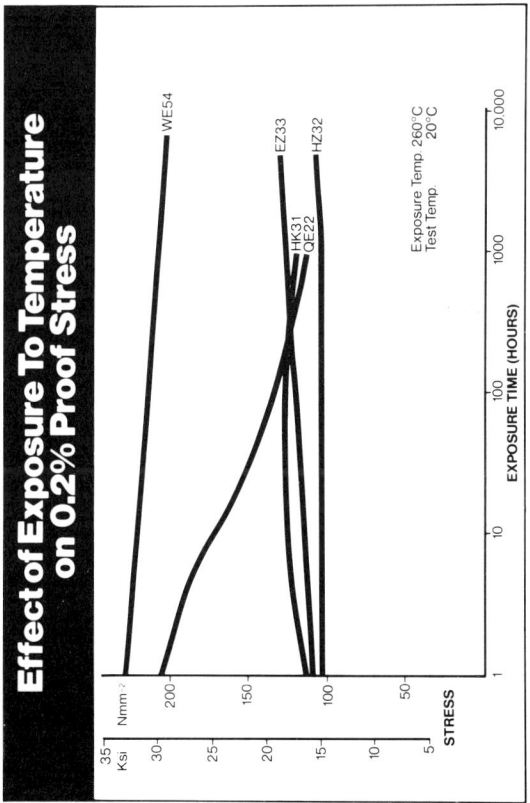

FIGURE 9

Stress/Time Relationship for 0.2% Total Strain at 260°C

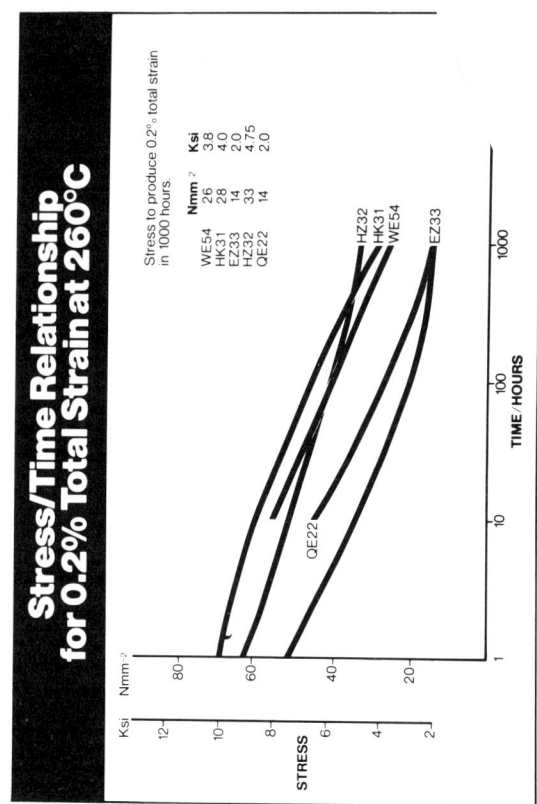

FIGURE 10

35

Wrought magnesium alloys

B KITTILSEN

The author is with Norsk Hydro a.s,
Porsgrunn, Norway.

SYNOPSIS

Magnesium wrought products have been produced for
more than 50 years, mainly used in areas where its
light weight is of prime importance.

In order to increase the use of wrought
magnesium the production costs have to be reduced
and the properties have to be improved.

The paper describes improvements in d.c.
casting technology, and comparison of indirect and
direct extrusion of some magnesium alloys.

The paper also gives the properties of a new
wrought magnesium alloy and experimental magnesium
composite extrusions.

WROUGHT MAGNESIUM ALLOYS

I am going to take the words of Dr. Willard Dow
as my own.

"When seeking out new potential uses for
magnesium", he used to say, "look for anything
that has to be pushed, pulled, picked up, carried,
or otherwise moved".

If we look at applications of wrought magnesium
alloys after World War II this statement certainly
has been followed up.

Magnesium has been labelled "the world's
lightest structural metal". Therefore, it is mainly
used in areas where its light weight is of prime
importance.

Wrought magnesium has been widely utilized in
transportation, handling equipment and sport
equipment. Examples are truck bodies, hand trucks,
platform trucks, dock boards, utility carts, tools
for handling concrete, tennis rackets, plus
many others.

Available market data indicates a decline in
the use of wrought magnesium from 1972. Today's
production is about 7000 t.p.y. which is less than
half of the production in 1972.

Why did these products almost disappear from
the market? The serviceability was recorded to be
good, and it was strongly indicated by the people
involved, that "there are no reasons, except for
cost, that exclude magnesium from being used in
such products".

The competitiveness of magnesium, however,
will improve if mechanical properties can
be increased.

Recent efforts have been made to reduce
production costs and improve properties.

Traditional and possible prospective methods
of producing wrought magnesium products are shown
in Figure 1.

PRIMARY PRODUCTION, MELTING AND ALLOYING

New magnesium capacities and modernizing programs
for old plants contribute to make valuable cost
reductions in the production of primary magnesium.

Reduction in labour and energy costs represent
high cost reduction potentials. Recent developments
at Norsk Hydro have also significantly reduced both
labour and energy costs. This was achieved by using
an advanced method producing anhydrous magnesium
chloride and large advanced design electrolytic cells.

Cost reductions and quality improvements can
also be obtained if attention is paid to melting
and alloying techniques. Metal losses have to be
reduced, energy efficiency has to be improved, and
scrap has to be recycled.

New furnace technology has improved energy
efficiency and metal losses, and work is now being
carried out in order to modify the furnace for
scrap processing.

DEVELOPMENTS IN PRODUCTION OF EXTRUSIONS

Casting of billets

The technology for casting of billets has until
recently been represented by continuous and semi-
continuous d.c. casting machines with single- and
double strand casting tables.

Comparing this with modern multiple-strand
machines for aluminium, allowing 100 logs to be
cast at a single pouring, magnesium is losing with
regard to both quality and cost.

Casting speed is limited by those metallurgical
factors which cause centre cracks. Casting single-
strand or multi-strand has no influence on speed,
and the number of operators required is almost
the same. Therefore, production costs can be
significantly reduced by increasing capacity through
multi-strand casting and increased drop length.

Through development of hot top casting
technology for magnesium and modification of the
new Showa casting process (1) (Figure 2), Norsk
Hydro has made multi-strand casting of magnesium
possible. Problems related to refractory materials,
gas mixtures, safety and gas protection had
to be solved.

Pressurized gas is introduced into the mould through a narrow gap between the header and the mould. This lowers the contact point of the hot metal to the inner surface of the mould and reduces the contact area.

Thus, a very smooth cast surfacé is obtained (Figure 3).

The mould height is much shorter than in conventional d.c. casting processes. This is important as it improves both the quality of the billet and the casting speed.

Figure 4 shows the effect of short moulds, which in fact means increased secondary cooling and decreased primary cooling (cooling through the mould wall). The results, beside a very smooth billet surface, are a fine grained homogeneous structure, and very little surface segregation. Figure 5 shows that the surface segregation layer in AZ-31 alloy is only one tenth that of conventional d.c. cast billets.

The improved billet quality results in better extrusion economy, due to avoidance of scalping and higher extrusion speed.

An eight strand pilot casting machine based on the modified Showa-process will be in production later this year.

EXTRUSION

Extrusion of magnesium is very similar to the extrusion of aluminium. The same general principles apply to alloys of either metal. Extrusion speeds, however, are somewhat lower than for comparable aluminium alloys, limited by hot shortness tendencies and the mechanical properties required.

For cost reasons it is important to increase extrusion speed without sacrificing mechanical properties.

In order to achieve this, the temperature rise during extrusion should be kept as low as possible. This can be done if:

- indirect extrusion is used
- extrusion die is cooled
- billet temperature is lowered
- billet length is reduced
- billet is homogenized.

In the indirect extrusion process (Figure 6) there is no relative displacement between billet surface and the container wall. This can be of advantage to magnesium because of the fact that the pressure to overcome friction and shear forces in the container can be eliminated. From this it seems logical that the exit temperature of the extrusions will be lower.

Figure 7 shows a typical progress when magnesium is extruded in a direct and in an indirect extrusion press. The total pressure required in indirect extrusion is almost equal to the deformation pressure.

COMPARISON OF INDIRECT AND DIRECT EXTRUSION

Norsk Hydro is now running a project with the scope of characterizing the extrudability of different magnesium alloys by direct and indirect extrusion, in order to get qualitative relations between process parameters and mechanical properties (3). The alloys AZ-31, AZ-61, ZCM-711 and ZK-30 (Table 1) are included in the program.

A small 500 t vertical extrusion press was used for the initial part of the project. The same press was used for both indirect and direct extrusion. A container diameter of 75 mm and an extrusion ratio of 40:1 were chosen. The product was a rectangular bar (24.6 x 4.6 mm). Extrusion

speed and billet temperature were varied, and parameters like exit temperature, extrusion pressure, extrusion speed and container temperature were continuously recorded for each run.

The preliminary results are interesting. As indicated in Table 2 the exit temperature of the extrusions develops differently in indirect and direct extrusion. For the highest billet temperature, the exit temperature is highest in indirect extrusion for all the alloys tested. It is difficult to find a logical explanation for this. One possibility, however, could be that the flow pattern changes in the container as the billet temperature increases.

HOT SHORTNESS very often limits the extrusion speed.

Figure 8 shows how extrusion speed and billet temperature affects hot shortness of AZ-61 alloy. As the curves indicate, indirect extrusion allows the highest extrusion speed. When we also know that the exit temperature is higher in indirect extrusion, we can conclude that hot shortness is not only a result of high temperature, but a combination of temperature and flow.

In extrusion the metal flows faster in the centre than near the surface. This results in stress. Due to the different flow characteristics in indirect and direct extrusion, the stress is highest in direct extrusion. Therefore, direct extruded metal has a greater tendency to hot shortness.

The extrusion speed of AZ-61 can be doubled by choosing indirect extrusion instead of direct extrusion. This is also the case with the ZCM alloy, while the AZ-31 and the ZK-30 alloys extrude at almost the same speed in both processes.

MECHANICAL PROPERTIES depend on alloy, billet temperature and extrusion speed. As the billet temperature and the extrusion speed increase, the mechanical properties decrease with the exception of the ZCM alloy. Figure 9 shows how the tensile yield strength of the alloy ZK-30 is reduced as the billet temperature and the extrusion speed are increased.

The figure also shows that direct extrusion gives higher tensile properties than indirect extrusion when the extrusion conditions are the same. This effect was also found for the other test alloys, but in the case of the AZ-31 alloy, the difference in properties was smaller.

Experimental test runs are planned on large extrusion presses in order to check whether these characteristics are reproduced, relative to the small extrusion press used for the trials.

There are, however, reasons to believe that the results are correct because the same characteristic behaviour is also found for aluminium alloys by G. Lang (4).

Figure 10 shows that the hardness changes very little with changes in extrusion parameters, but the same tendencies can be observed as for the tensile strength.

The better mechanical properties of extrusions obtained at low billet temperature and low extrusion speed are caused by less tendency to grain growth after extrusion (Figure 11).

Using direct extrusion of AZ-61 alloy as an example we can see that at the same billet temperature, the lowest speed gives the finest grains, and at equal speed, the lowest billet temperature gives the finest grains.

The relationships between the exit temperature and the grain growth and the grain size and mechanical properties are basic metallurgical knowledge.

Therefore, this may explain why the best properties are obtained in direct extrusion.

CONTINUOUS EXTRUSION

A future generation of extrusion technology will probably include the continuous extrusion processes.

One commercial consolidation process is available today - the Conform process (Figure 12). The Conform machine has an extrusion chamber comprising a three sided groove in a rotating wheel and a stationary shoe which forms the fourth side of the chamber. As the wheel rotates, the feed material is gripped in the groove and advances towards the die which is supported in the groove. The shoe reduces the cross section of the groove. Hence the feed stock, which can be powder, is compacted and starts to extrude.

Extremely high extrusion speeds have been recorded for aluminium, but the product size is still a limitation.

Various types of aluminium; fine powder, granules, chopped scrap and swarf have been successfully extruded into profiles. Even molten aluminium alloys have been used for feeding the Conform machine (5).

Magnesium granules have been extruded into rod. Except for minor heating problems in the starting phase of the Conform extrusion, the result was encouraging.

Further development is waiting for an efficient method to produce rapidly solidified magnesium powder.

Having an alloy feedstock of fine grain size is a good basis for obtaining a high strength product with adequate ductility and toughness.

Another interesting feature is the possibility of extruding premixed metal powders which would open a new world of alloys.

It is also possible to see a future application of the Conform machine in the production of metal matrix composite extrusions; mixtures of metal powder and chopped fibres seem to be a suitable feed material.

ALLOY DEVELOPMENT

Magnesium wrought alloys are split into three groups, those containing manganese, those containing aluminium, and those containing zirconium as the key alloying element.

Binary magnesium manganese alloys provide only medium strength, but have been used to some extent due to their good extrudability, good weldability and good corrosion properties.

A wider use, however, has been made of aluminium as the major alloying element, and a range of useful alloys has been produced.

Less use has been made of zinc, despite the fact that zinc is an excellent solid solution strengthener. In the absence of an effective grain refiner such as zirconium the alloys show increasing hot shortness with decreased weldability as the zinc content increases and the maximum amount of zinc that can be tolerated is limited. While zirconium additions reduce hot shortness they do not eliminate the lack of weldability with zinc contents above about 3%.

Zr is not only an efficient grain refiner, it also significantly increases the recrystallisation temperature from 100°C to 400°C in the magnesium zinc alloys.

One medium strength magnesium zinc alloy without zirconium, however, offers economic and to some extent also technical advantages over the present commercial wrought magnesium alloy, AZ-31. That is ZM-21, containing 2% zinc and 1% manganese. The ZM-21 alloy is readily hot workable due to suppression of the zinc rich second phase which normally restricts plastic deformation.

Magnesium Elektron Limited (MEL) has found that the addition of copper to the magnesium-zinc alloys results in a wrought alloy which responds to ageing and full heat treatment (6). In the latter condition yield strength higher than those of currently used wrought magnesium alloys was found (Table 3).

MEL developed an alloy, ZCM-711, which is weldable in both as-extruded and T-6 condition.

With a zinc content of 6 - 7% the tensile properties are reduced when the copper content increases above 1%. On the other hand, castability of the alloy is improved with increasing copper additions and, of particular importance, the solidus temperature of the alloy increases with increasing copper content permitting the solution heat treatment temperature to be raised to ensure maximum solution of zinc.

The copper content in the final alloy therefore, is based on a compromise.

Extrusions have been carried out with billet temperatures from 330°C to 400°C. The results show that the higher temperature gives the better properties even in the as-extruded condition. This is contrary to experience with other magnesium alloys where the best properties are achieved by extruding cold and slow.

The extrusion speed is reported to be higher than for the alloy AZ-61, but somewhat lower than for the AZ-31 alloy.

The corrosion resistance of the ZCM alloy is somewhat lower in the as-extruded condition than AZ-31 B, although the corrosion properties improve on full heat treatment.

MAGNESIUM MATRIX COMPOSITES

Increased demand for higher performance has stimulated interest in reinforced metal matrix composites. Magnesium, with its low density, is extremely suitable as a matrix material.

Most of the work is being carried out in the casting field, but some work on extrusion of magnesium matrix composites with silicon carbide whiskers is also creating interest (7) (Table 4). The mechanical properties of these extrusions are reported to be encouraging. Published data shows a yield strength up to 425 MPa and an elastic modulus of more than 100 GPa.

The extrusion billets were produced by blending magnesium powder and SiC fibres, consolidating it by compactions, and then homogenizing the billets prior to extrusion (Figure 13).

Until efficient production methods for rapidly solidified magnesium powders are developed and continuous extrusion processes can extrude complicated, large shapes, the production of composite magnesium billets through normal d.c. casting methods should find a solution.

SHEETS AND PLATES

Production of magnesium sheet and plate are high cost operations. If rolled magnesium is to become more competitive with aluminium, metal yield and economics must improve considerably.

Further development will be based on continuous strip casting in order to avoid costly breaking down of slabs.

While the production figures are low today, encouraging short term developments are helping to give renewed interest in magnesium wrought products.

REFERENCES

1 S. Yanagimoto and R. Mitamura,
   "Application of New Hot Top Process to
   Production of Extrusion and Forging Billet".
   Third International Aluminium Extrusion
   Technology Seminar, April 1984, Atlanta, USA.

2 B. Kittilsen, "How to Expand the Market for
   Magnesium Wrought Products".
   The 42nd Annual World Magnesium Conference,
   May 1985, New York.

3 H. Gjestland, "Extrusion of Magnesium".
   The fourth Scandinavian Symposium on
   Materials Science, August 1986,
   Trondheim, Norway.

4 G. Lang, "Strangpressen beim direkten und
   indirekten Pressen von Al 99.6 und ihre
   Abhängigkeit vom Pressverhältniss (I, II)".
   Aluminium no. 8, 9, 1985.

5 B. Maddock, "GIESSPRESSEN von Profilen −
   eine Weiterentwicklung des Conform-Verfahrens".
   Aluminium no. 6, 1985.

6 W. Unsworth, "A High Strength Wrought Magnesium
   Alloy". The 39th Annual World Magnesium
   Conference, May 1982.

7 H. J. Rack and P. W. Niskanen,
   "Extrusion of Discontinuous Metal Matrix
   Composites". Light Metal Age, February 1984.

TABLE 1   COMPOSITION OF ALLOYS TESTED

	Zn	Al	Zr	Mn	Cu	Mg
AZ31	0.84	3.0		0.54		Rem
AZ61	0.99	6.9		0.37		Rem
ZCM	6.4			0.82	1.18	Rem
ZK30	3.0		0.65			Rem

TABLE 2   EXIT TEMPERATURE ON EXTRUSIONS

Billet Temperature	Exit Temperature Direct Extrusion	Exit Temperature Indirect Extrusion
High (400°C)	Low	High
Low (300°C)	High	Low

TABLE 3   MAGNESIUM ALLOY ZCM-711

Chemical Composition

Zn (%)	Cu (%)	Mn (%)	Mg (%)
6.0-7.0	1.0-1.5	0.5-1.0	REM

Typical Tensile Properties *

Condition	TYS (MPa)	TS (MPa)	Elongation (%)
F	183-194	285-296	10-13
T-5	244-270	311-326	6-10
T-6	347-357	367-382	4-10

* Extruded bar 5/8 inch diameter

TABLE 4   MECHANICAL PROPERTIES OF EXTRUDED AZ-31B ALLOY AND EXTRUDED Mg-SiC COMPOSITE BILLETS

Property		Material		
		AZ-31B	Mg+10%SiC	Mg+20%Sic
TYS	(MPa)	200	321	425
TS	(MPa)	260	375	456
Elongation	(%)	15	1.6	0.9
Elastic Modulus	(GPa)	45	70	102

40

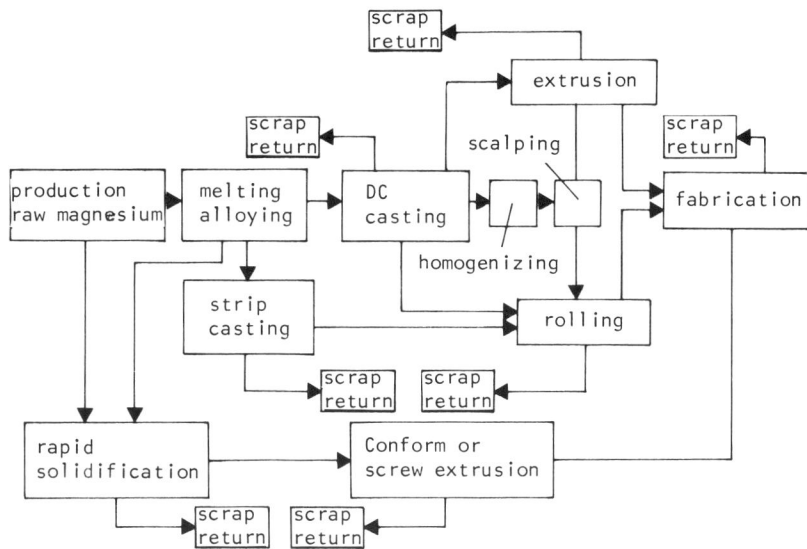

Figure 1   Production of wrought magnesium products

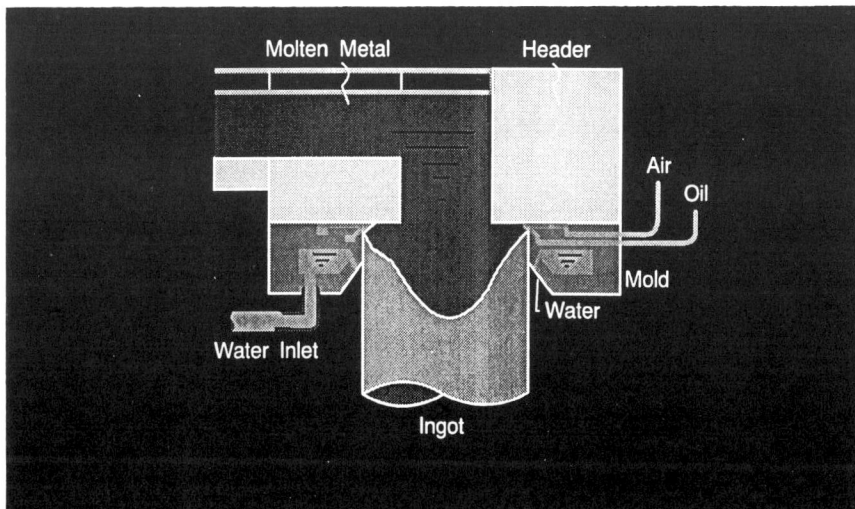

Figure 2   The Showa DC casting process

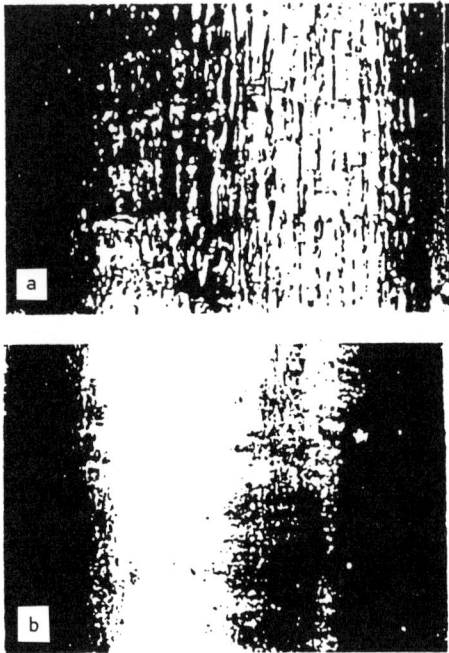

Figure 3   Surface of cast AZ-31 alloy billet:
            (a) cast without gas pressure, (b)
            cast with gas pressure (modified
            Showa process)

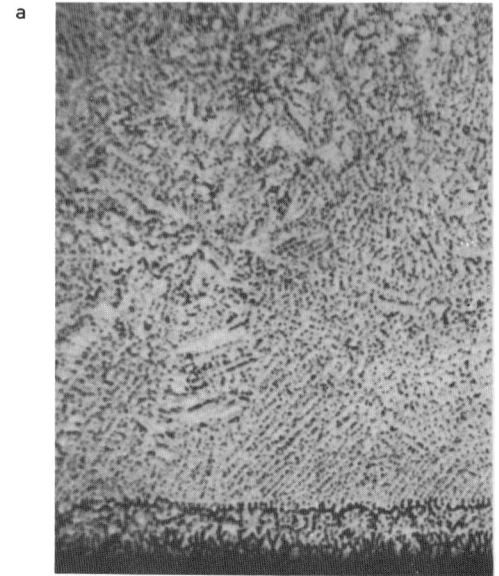

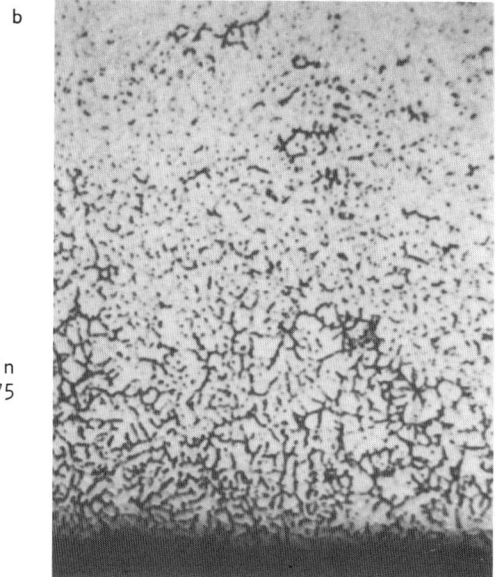

Figure 5   Depth of surface segregation in
            DC cast AZ-31   x 75
            (a) new technology
            (b) conventional technology

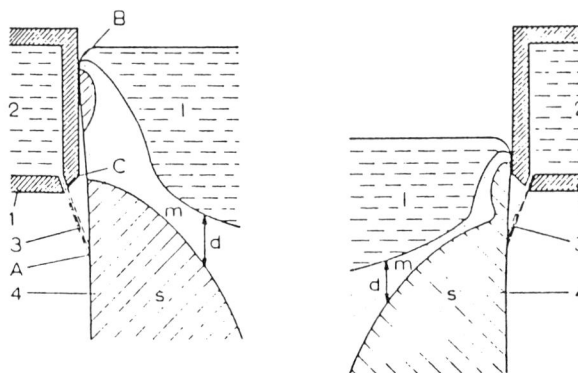

Effect of level of metal in the mold on solidification conditions
left: metal level high, right: metal level low; 1 Mold, 2 Water chamber, 3 Water jet,
4 Strand being cast; l liquid, m mushy, s solid

Figure 4   Effect of metal level in the mould
            on solidification conditions

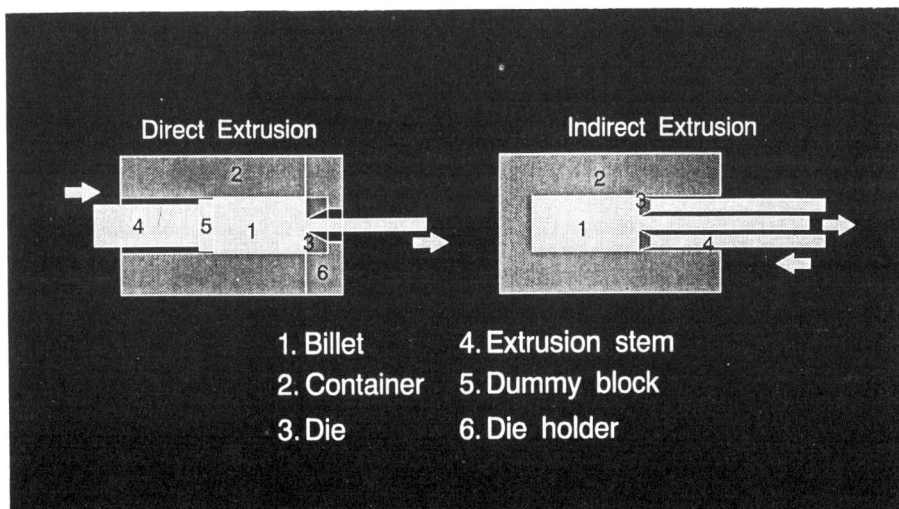

Figure 6   Basic methods of extrusion

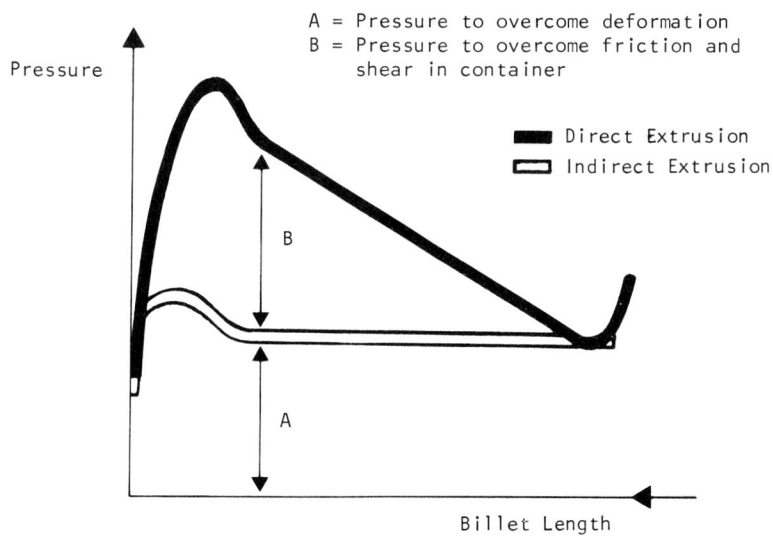

Figure 7   Pressure required during direct and
            indirect extrusion

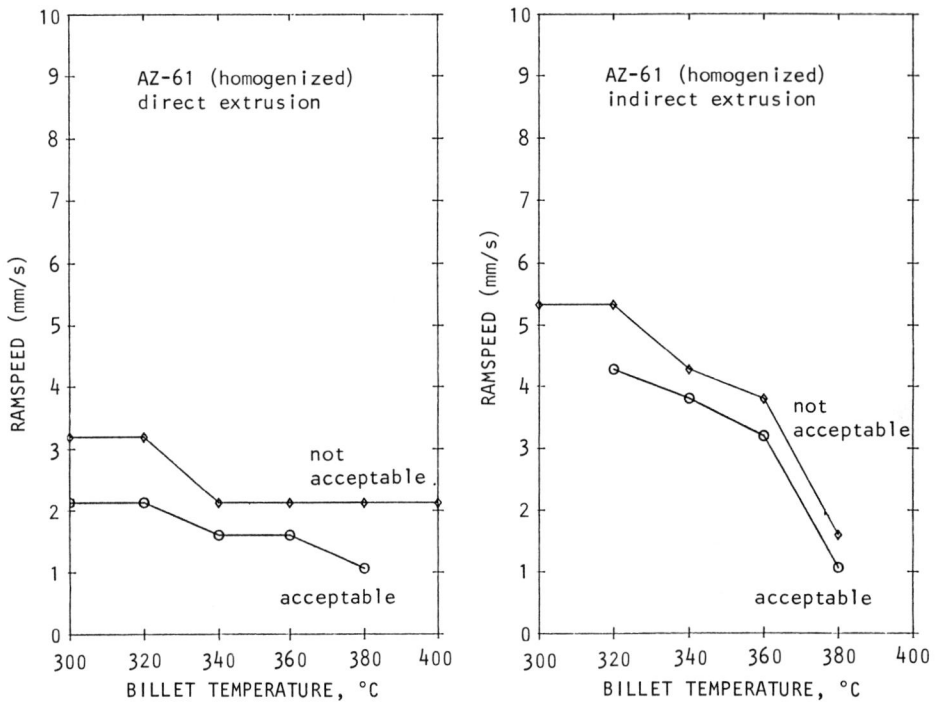

Figure 8   Hot shortness at different billet temperatures
and extrusion speeds

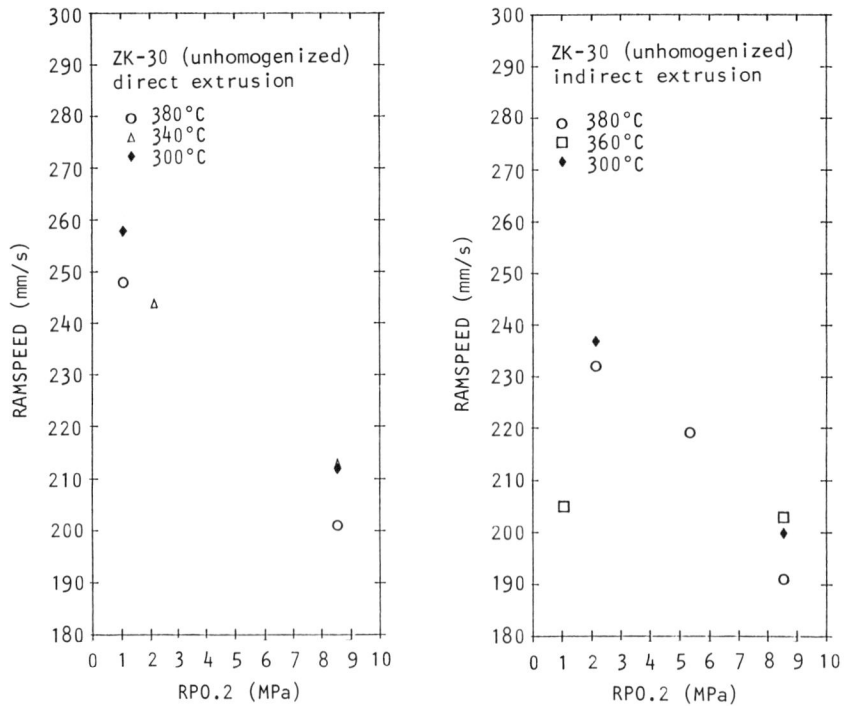

Figure 9   Tensile yield strength versus extrusion speed

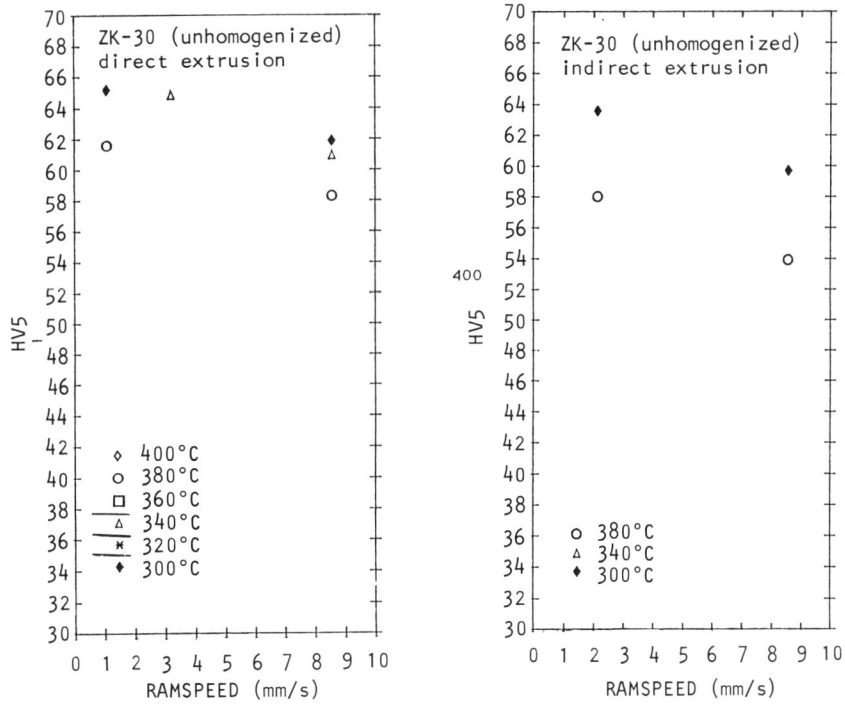

Figure 10   Hardness versus extrusion speed

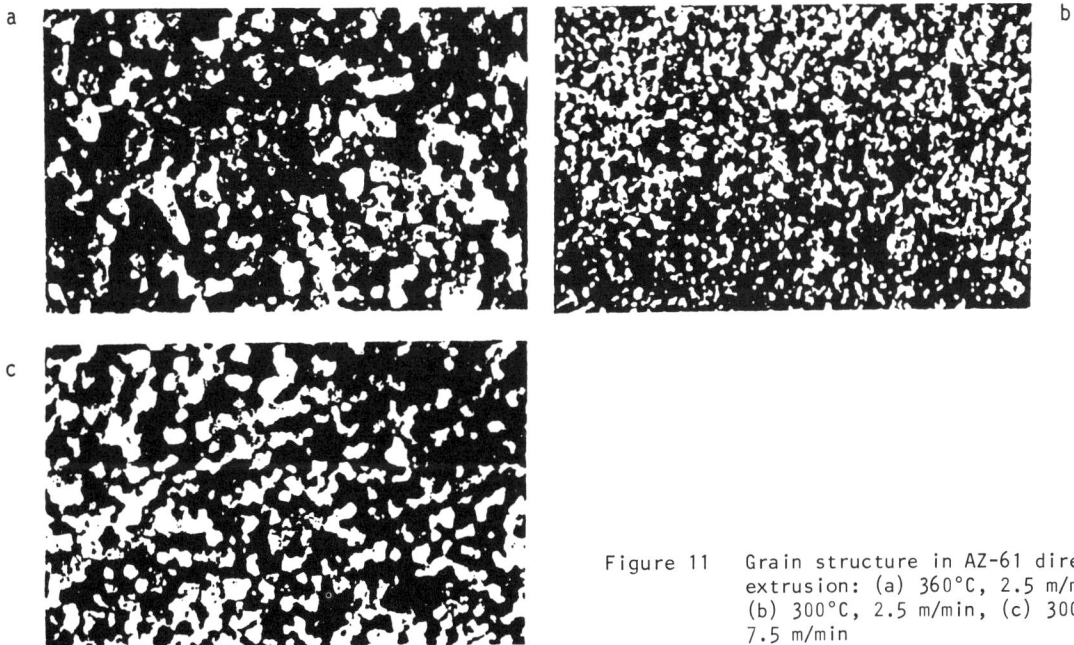

Figure 11   Grain structure in AZ-61 direct extrusion: (a) 360°C, 2.5 m/min, (b) 300°C, 2.5 m/min, (c) 300°C, 7.5 m/min

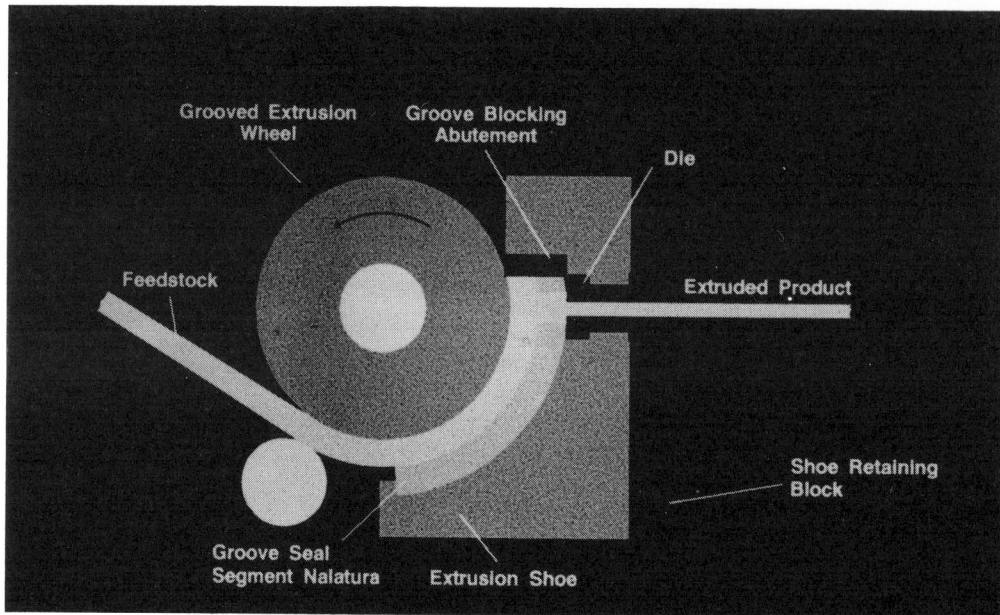

Figure 12   Conform extrusion

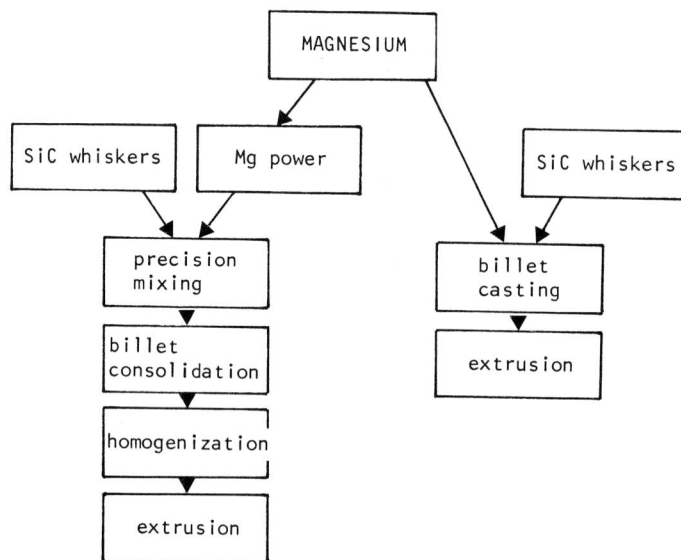

Figure 13   Production of SiC reinforced magnesium
matrix composite extrusions

Structure–property relationships in cast magnesium alloys

G W LORIMER

The author is in the Department of Metallurgy and Materials Science, University of Manchester (UMIST), Manchester, England.

SYNOPSIS

The structure-property relationships in a number of magnesium casting alloys are discussed, with specific reference to heat treatable alloys. The background phase diagrams and the effect of systematic variations in alloy content on properties, particularly in binary magnesium-X systems, have been discussed in detail by Emley[1] and the reader is referred to this excellent text. The precipitation phenomena in magnesium alloys have recently been reviewed by Polmear[2]. In the present paper, for those alloys where new developments have not occurred since Polmear's book was published, the precipitation sequence is merely referred to for completeness. Where significant alloy development has taken place these results are discussed in some detail. The author has also found it useful to refer to the review papers by Stratford[3] and Waibel[4].

GRAIN REFINEMENT

To produce optimum tensile properties and adequate toughness it is imperative to obtain effective grain refinement during casting. In most magnesium casting alloys, grain refinement is accomplished by the addition of 0.6 to 0.7wt% Zr, usually in the form of a fluorzirconate, to the melt. The molten magnesium reduces the fluorzirconate and liberates zirconium. The exact mechanism by which the zirconium produces grain refinement is not clear. Certainly zirconium is concentrated at the centres of grains in cast microstructures. As pointed out by Polmear[2], the lattice parameters of α-zirconium (a = 0.323 nm, c = 0.514 nm) are similar to those of magnesium (a = 0.320 nm, c = 0.520 nm); if zirconium particles precipitated first from the liquid they could act as effective nucleating agents. Alternatively a zirconium compound may be formed.

The most widely used magnesium die casting alloys contain, typically, 6 to 10wt% aluminium, which forms intermetallic compounds with liquid zirconium. Thus zirconium cannot be used for grain refinement. Magnesium-aluminium alloys can be grain refined either by superheating to 850°C and cooling quickly to the normal casting temperature or by the addition of carbon-containing compounds, such as calcium carbide or hexachloroethane to the melt. The nucleation of magnesium grains is assumed to occur on Al_9C_3 or a more complex Al,C compound.

MAGNESIUM-ALUMINIUM

The maximum solid solubility of aluminium in magnesium is 12.7wt% at 437°C, which decreases to ~2wt% at 100°C. Thus it might be expected that with the appropriate homogenization, quenching and ageing treatments a significant ageing response would result. However during ageing a coarse, heterogeneous distribution of the equilibrium $Mg_{17}Al_{12}$ phase is precipitated; no zones or intermediate precipitates have been reported and the ageing response is poor. Clark[5] observed that direct nucleation of the equilibrium $Mg_{17}Al_{12}$ phase occurred as platelets on $\langle 0001 \rangle_{Mg}$, with preferential heterogeneous nucleation on twin boundaries. The $Mg_{17}Al_{12}$ interparticle distance was large, which accounted for the poor age hardening response.

An additional complication in magnesium-aluminium alloys is the presence of a competitive cellular reaction which nucleates at grain boundaries. The cellular reaction predominates in slowly cooled alloys and competes with the matrix precipitation of $Mg_{17}Al_{12}$ during ageing.

The variation in the tensile properties of magnesium-aluminium alloys as a function of aluminium concentration have been summarized by Fox[6]. In the solution-treated condition U.T.S. values up to 240 Nmm^{-2} and an elongation of 10%, can be obtained for an alloy containing 10wt% aluminium, but the 0.2% P.S. is low, 70 N mm^{-2}. In the fully heat-treated condition the U.T.S. values are unchanged, the 0.2% P.S. is increased to 100 N mm^{-2}, but the ductility is dramatically reduced. Acceptable ductility values can only be obtained if the aluminium content is kept below 8wt%.

Most commercial magnesium-aluminium alloys also contain zinc, e.g. AZ63 and AZ91, where high zinc contents are usually associated with low aluminium contents, and vice versa, to give a total aluminium plus zinc concentration of about 10wt%. The addition of zinc to a magnesium-aluminium alloy results in a moderate increase in strength. Mg-Zn-Al alloys are used almost exclusively as die casting alloys in the as-cast condition.

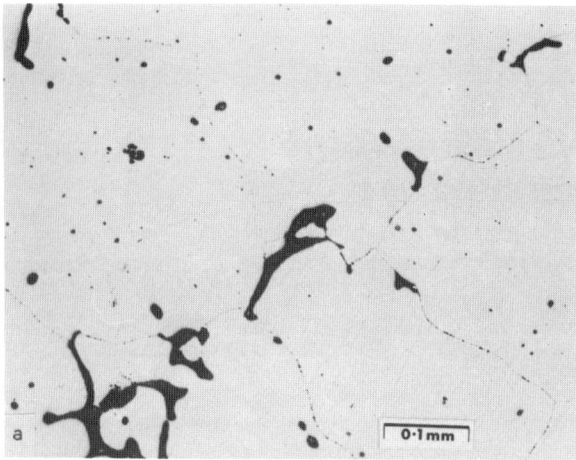

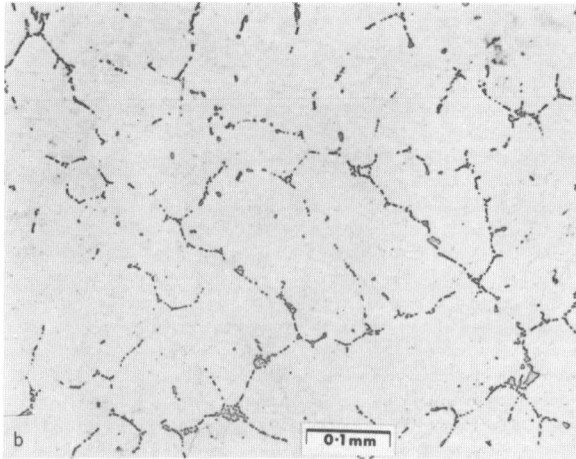

Figure 1. Effect of the addition of copper on
the morphology of the eutectic in a
Mg-6wt%Zn alloy. (a) Z6 cast,
solution treated 8 hrs at 330°C.
(b) 2C61½ cast, solution treated
8 hrs at 435°C.

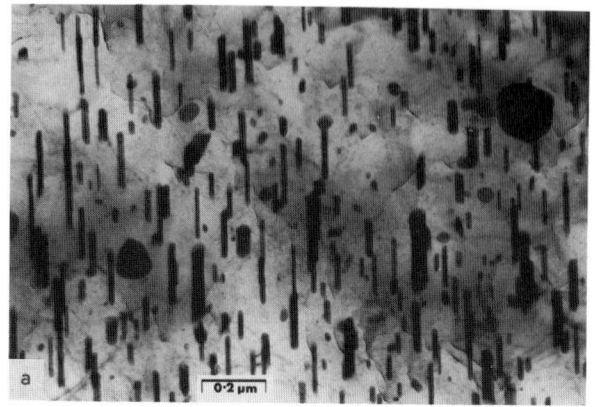

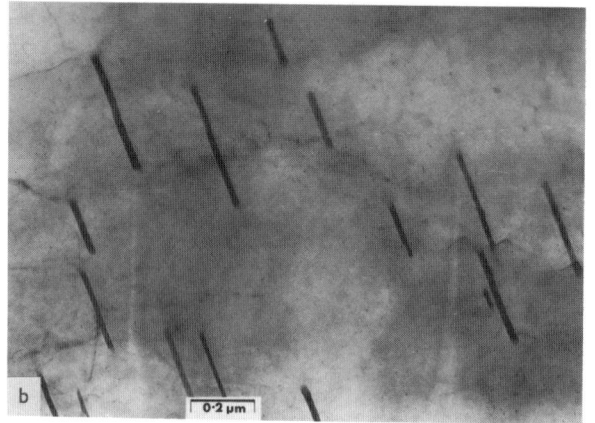

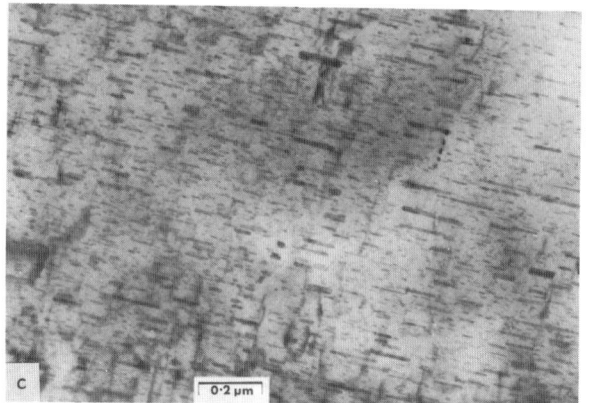

Figure 2. (a) Z6 aged 8 hours 200°C showing a
moderate precipitate density
(b) Z6 as in (a) but a grain with a
low precipitate density
(c) ZC61½ aged hours at 200°C
showing a precipitate density similar
to that observed in all grains.

MAGNESIUM-ZINC

Magnesium-zinc binary alloys are usually grain refined by the addition of zirconium. Although the zirconium addition reduces the ageing response of the binary alloy this is more than compensated for by the increase in strength associated with grain refinement, e.g. ZK51 and ZK61. Both alloys have good strength: U.T.S. values of 205 and 310 N mm^{-2}, and adequate ductility, 3.5 and 10% elongation, in the T5 and T6 conditions, respectively. The commercial use of Mg-Zn-Zr alloys is limited as they are more susceptible to microporosity and hot cracking and less weldable than Mg-Zn-Al alloys.

The precipitate sequence in binary Mg-Zn alloys has been studied by a number of authors, and conflicting results are reported. Sturkey and Clark[7] and Clark[8] investigated alloys containing 4 to 8wt% Zn in the temperature range 95 to 200°C and identified the precipitation sequence as

$$Mg_{ssss} \rightarrow \beta' \ (MgZn_2) \rightarrow MgZn$$

where the β' phase formed as needles parallel to $\langle 0001 \rangle_{Mg}$. Work by Murakami et al.[9] confirmed the results of Clark[8] in the temperature range 150 to 200°C, but between 70 and 100°C they reported the formation of plate-shaped G.P. zones on $\langle 10\bar{1}1 \rangle_{Mg}$ as a precursor to β'.

Gallot[10] used X-ray diffraction to investigate precipitation in a Mg-6wt%Zn alloy aged at temperatures between 50 and 250°C. He identified a second transition precipitate, β_2', and the equilibrium phase as Mg_2Zn_3. He proposed the precipitation sequence

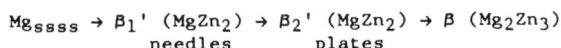

$$Mg_{ssss} \rightarrow \underset{\text{needles}}{\beta_1' \ (MgZn_2)} \rightarrow \underset{\text{plates}}{\beta_2' \ (MgZn_2)} \rightarrow \beta \ (Mg_2Zn_3)$$

Takahashi, Kojima and Takanishi[11] also used X-ray diffraction to study precipitation in a magnesium 3.6wt% Zn alloy aged at temperatures between room temperature and 140°C. They detected two types of G.P. zones as well as two transition precipitates and gave the precipitation sequence as

$$Mg_{ssss} \rightarrow \underset{\text{on } \langle 11\bar{2}0 \rangle_{Mg}}{G.P._{-}1} \rightarrow \underset{\text{on } \langle 0001 \rangle_{Mg}}{G.P.2} \rightarrow \beta_1' \rightarrow \beta_2'$$

The strengthening mechanisms in precipitation hardened magnesium zinc alloys have been studied in wrought material, but the discussions are also valid for cast alloys. Two schools of thought have emerged. Clark[8] proposed that the main strengthening at peak hardness was due to Orowan looping because the transitional precipitate rods (β_1' or β_2') were too widely spaced to necessitate shearing of the precipitates. Chun and Byrne[12,13] postulated that particle shear was the main strengthening mechanism whether β_1', β_2' or β precipitates were present. Mima and Tanaka[14] proposed that cutting occurred in the under- and peak-aged conditions, but in the over-aged condition Orowan looping predominated.

MAGNESIUM-ZINC-COPPER

The addition of small amounts of copper, up to 3 wt%, to binary magnesium-6wt% zinc alloys eliminates the tendency of the binary alloy to tear and crack and reduces the susceptibility of the alloy to microporosity. In addition the alloy can be effectively grain refined by bubbling $C_2F_2Cl_2$ through the melt. The details of the development of the alloy are given in the paper by Unsworth[15].

Moss[16] made a systematic investigation of the age hardening response at 150, 200 and 250°C of three alloys: Z6, ZC61½ (Mg-5.6wt% Zn-1.4wt% Cu), and ZC63 (Mg-5.9wt% Zn-3.2wt% Cu). The effect of the copper addition was to change the morphology of the eutectic phase from divorced eutectic in Z6 to a lamellar intergrowth of magnesium and intermetallics (Fig. 1). This change in morphology and the grain refinement resulted in an increase in the ductility of the cast copper-containing alloys, as compared with the binary magnesium-zinc alloy. The ageing response of the three alloys was monitored at 150, 200 and 250°C; that of ZC61½ was found to be superior to that of Z6 and ZC63. The precipitation sequence in ZC61½ and ZC63 was investigated using transmission electron microscopy and found to be the same as in Z6:

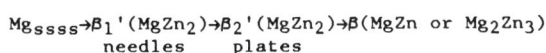

$$Mg_{ssss} \rightarrow \underset{\text{needles}}{\beta_1' \ (MgZn_2)} \rightarrow \underset{\text{plates}}{\beta_2' \ (MgZn_2)} \rightarrow \beta (MgZn \text{ or } Mg_2Zn_3)$$

G.P. zones were not detected at the temperatures investigated.

The addition of copper to the binary Z6 alloy not only changed the morphology of the eutectic, but copper was incorporated in the eutectic phase as $Mg(Cu,Zn)_2$. As copper was added to the Z6 alloy the volume fraction of eutectic phase in the as-cast alloy increased, but following solution heat-treatment the volume fraction of $Mg(Cu,Zn)_2$ was reduced. Also, it was possible to solution heat treat ZC61½ at 445°C as compared with 340°C for Z6. The effects of a higher solution treatment temperature and the replacement of some of the zinc in the grain boundary eutectic with copper released more zinc to form a larger volume fraction of the strengthening β_1' precipitates (Fig. 2). Moss[16] also found a wide variation in the density of β_1' from grain-to-grain in the Z6 alloy: some grains contained an order of magnitude fewer precipitates than another (Fig. 2a,b). This effect was not observed in ZC61½ or ZC63 (Fig. 2c). Thus, as well as releasing more zinc to form the β_1' and β_2' precipitates, the copper addition may have also catalyzed the nucleation of the β_1' and/or β_2' phases.

MAGNESIUM-RARE EARTH

Magnesium alloys containing rare earths form the basis of a number of commercial magnesium alloys, e.g. EZ33 and ZE41, which have good creep resistance up to 160°C. ZE63, Mg-5.8wt% Zn-2.5wt%Re-0.7wt%Zr can be precipitation hardened (to a finite depth) by diffusing hydrogen into the alloy at high temperatures. As-cast ZE63 contains a massive Zn,RE-rich grain boundary phase which produces a brittle alloy. The grain boundary phase can be eliminated in thin specimens or surface layers by high temperature treatment in hydrogen, during which the RE elements are precipitated as hydrides and the Zn released into solid solution. Not only is the brittle grain boundary phase eliminated but a high volume-fraction of hydride is precipitated, predominantly near the surface, and produces a hard, fatigue resistant, surface layer.

As part of the development of a commercial Mg-Y-RE alloy (Unsworth[15]), Karimzadeh[17] has carried out a study of the precipitation reaction

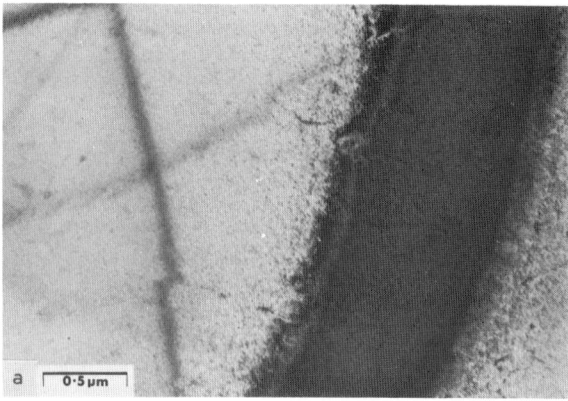

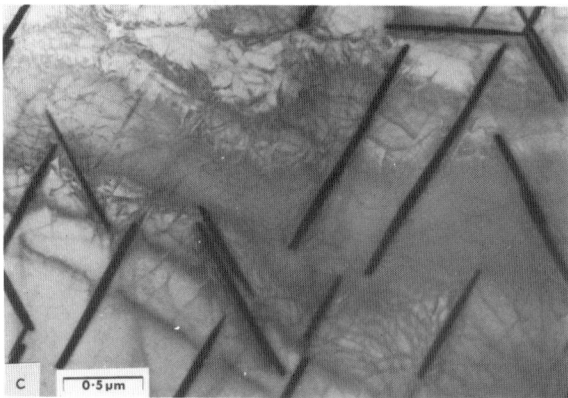

Figure 3.   Mg-5.48wt%Y-2.96wt%Nd-0.4wt%Zr
solution treated at 535°C, quenched
and (a) aged at 175°C for 240 hours
showing the β″ precipitates.
(b)   aged at 250°C for 10 hours
showing the β′ precipitates.
(c)   aged at 350°C for 3 hours showing
the equilibrium β precipitates.

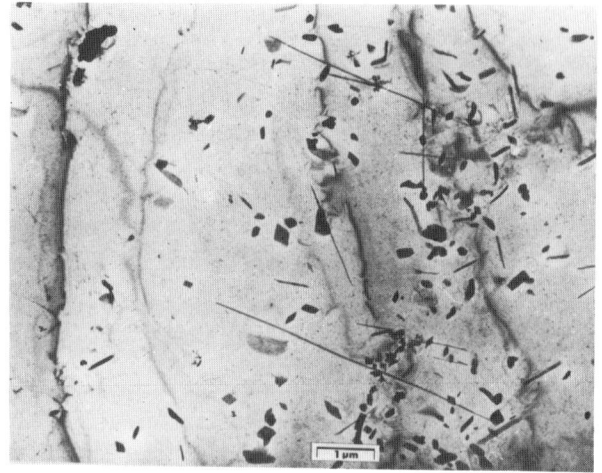

Figure 4.   QK91 (Mg-8.73Ag-0.51Zr) aged at 200°C
for 32 hours showing a heterogeneous
distribution of γ′ (from Gradwell[20]).

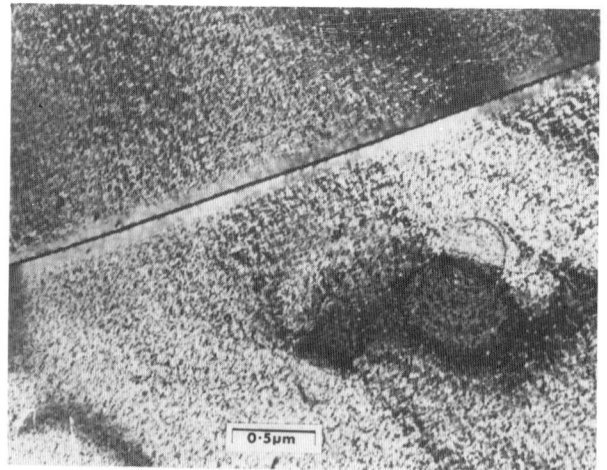

Figure 5.   QE22 (Mg-2.5wt%Ag-2.0wt%RE-0.6wt%Zr)
aged at 200°C for 4 hours showing
a high density of γ and β precipitates
(from Gradwell[20]).

in Mg-Nd, Mg-Y and Mg-Y-Nd alloys. In Mg-10wt% Y aged at temperatures between 200 and 320°C Karimzadeh[17] identified the precipitation sequence as

$$Mg_{ssss} \rightarrow \beta'' \rightarrow \beta' \rightarrow \beta(Mg_{24}Y_5)$$

The main strengthening precipitates were identified as the homogeneously distributed β'' phase, which formed as plates on the $\langle 10\bar{1}0 \rangle_{Mg}$. The β'' plates had a base-centred orthorhombic crystal structure with lattice parameters $a = 6.4$ Å, $b = 22.23$ Å and $c = 5.21$ Å. The β' precipitates were found to have the same crystal structure as the β'' phase but were larger and heterogeneously distributed; the hardening effect of the β' precipitates was small. The equilibrium β phase formed at 320°C and had little effect on the strength of the alloy.

The results of Karimzadeh[17] were basically in agreement with the observations of Mizer and Clark[18] and Mizer and Peters[19], although these authors identified the β'' and β' precipitates as having a simple monoclinic structure while Karimzadeh[17] indexed the diffraction patterns as from a base-centred orthorhombic cell. Mizer and Clark (1981) suggested that the equilibrium β precipitates were cubic while Karimzadeh found them to have a B.C.C. structure.

In the investigation of a Mg-3wt% Nd alloy aged at temperatures between 200 and 300°C Karimzadeh[17] found the precipitation sequence

$$Mg_{ssss} \rightarrow G.P. \text{ zones} \rightarrow \beta'' \rightarrow \beta' \rightarrow \beta(Mg_{12}Nd)$$

In the Nd alloy the β'' phase, the main strengthening precipitate, was found to have the DO_{19} superlattice with an "a" parameter twice that of the magnesium matrix and a "c" parameter equal to that of the matrix. The β' precipitates had a F.C.C. structure with $a = 7.35$ Å and a chemistry close to Mg_2Nd_{17}, and the equilibrium β phase had a B.C.C. structure with a chemistry close to $Mg_{12}Nd$. These results are in general agreement with the previous investigations by Gradwell[20] and Pike and Noble[21]. Karimzadeh found the crystal structure of the β' phase to be F.C.C., in agreement with the work of Gradwell[20], and not H.C.P. as proposed by Pike and Noble[21].

Karimzadeh[17] examined the ageing response and the precipitation sequence in two Mg-Y-RE-Zr alloys. One alloy, WE54X, based on the commercial WE54 alloy, contained Mg-6.85wt% Mischmetal (75%Y + 25% heavy rare earths)-1.82wt%Nd-0.52wt%Zr. The second alloy, made from pure yttrium, had the composition Mg-5.08wt% Y-2.96wt%Nd-0.4wt%Zr. The precipitation sequence observed in both alloys was the same but the ageing response of the alloy containing pure yttrium was slightly superior, at all temperatures, to the alloy containing Mischmetal.

At ageing temperatures between 175 and 350°C the precipitation sequence was identified as

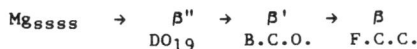

$$Mg_{ssss} \rightarrow \underset{DO_{19}}{\beta''} \rightarrow \underset{B.C.O.}{\beta'} \rightarrow \underset{F.C.C.}{\beta}$$

All the precipitates formed as plates on the $\langle 10\bar{1}0 \rangle_{Mg}$ planes. The β'' plates, which formed at 175 and 200°C, Fig. 3a, had the DO_{19} crystal structure, the same as β'' in the binary Mg-Nd alloy, with the orientation relationship

$$[0001]_{\beta''} || [0001]_{Mg}$$

and $\quad (01\bar{1}0)_{\beta''} || (0001)_{Mg}.$

The β' precipitates, Fig. 3b, which formed at temperatures between 200 and 250°C had a base-centred orthorhombic structure with the same orientation relationship and lattice parameters as the β' precipitates in the Mg-Y binary, i.e.

$$a = 6.4 \text{Å} \quad b = 22.23 \text{ Å and } c = 5.21 \text{ Å}$$

$$[001]_{\beta'} || [0001]_{Mg}$$

$$(100)_{\beta'} || (2\bar{1}\bar{1}0)_{Mg}.$$

The equilibrium β precipitates, Fig. 3c, were detected in the matrix after ageing at 300°C and above. They had an F.C.C. structure with a lattice parameter of $a = 22.23$ Å and an orientation relationship of

$$(011)_{\beta} || (0001)_{Mg}$$

$$[1\bar{1}1]_{\beta} || [1\bar{2}10]_{Mg}.$$

The commercial WE54 alloy has good castability, a 0.2% P.S. of 185 N mm^{-2} and a U.T.S. of 255 N mm^{-2} with a 2% elongation. It has excellent retention of strength and good creep resistance at temperatures up to 250°C. The tensile and creep properties of the alloy are discussed by Unsworth[15] and Karimzadeh et al.[22] in this volume. It is sufficient to note here that these authors report that, as the ageing temperature was increased, the 0.2% P.S. and U.T.S. at peak hardness decreased, and the % elongation increased. An interesting, and unexpected, observation made by these workers[22] was that the amount of intergranular (brittle) fracture apparently increased with the increase in % elongation. The excellent high temperature creep resistance of the alloy may be associated with the geometry of the β' and β precipitates, Fig. 3(b) and (c). These plate-shaped precipitates form an interlocking network of precipitates which will be highly resistant to coarsening and may be responsible for the observed microstructural stability during creep deformation.

MAGNESIUM-THORIUM

The age hardening response in binary Mg-Th alloys was identified by Kent and Kelly[23] and Mushovic[24] as

$$Mg_{ssss} \rightarrow \beta'' \rightarrow \beta' \rightarrow \beta(Mg_{23}Th_6).$$

The β'' phase had the DO_{19} structure and formed as plates on $\langle 10\bar{1}0 \rangle_{Mg}$, the β' phase was F.C.C. with $a = 8.59$ Å and the β phase was the equilibrium F.C.C. $Mg_{23}Th_6$. Stratford[25] has suggested that, in the presence of Zr, the β'' phase can transform in-situ to produce a coherent, hexagonal phase with $a = 10.66$ Å and $c = 8.28$ Å. Stratford proposed that subsequently this phase transformed into two different phases, one plate-like and parallel to $\langle 10\bar{1}0 \rangle_{Mg}$ and the other disk-like and parallel to $\langle 0001 \rangle_{Mg}$. Both of these precipitates then transformed into the equilibrium phase $Mg_{23}Th_6$.

The precipitation sequence in Mg-Th alloys, both in the binary and zirconium-containing ternary, must still be considered as speculative, particularly the role of each precipitate in nucleating succesive precipitate phases.

Alloys based on the Mg-Th system have many similar characteristics to the Mg-RE alloys.

Both are the basis of commercial derivatives which have excellent high temperature creep properties, good castability and are weldable. HK31A (Mg-3.3Th-0.7Zr) in the T6 condition and HZ32A (Mg-3.3Th-2.1Zn-0.7Zr) in the T5 condition have been used[4] up to temperatures of 350°C. The good creep resistance of the thorium-containing alloys may have similar origins to that of the Mg-Y-RE-Zr alloys: interlocking, plate-shaped precipitates on $\langle 10\bar{1}0 \rangle_{Mg}$ in an array that is highly resistant to particle coarsening.

MAGNESIUM-SILVER

The Mg-Al binary system exhibits an extensive solid solubility of silver in magnesium, 15.3wt% at the eutectic temperature of 472°C, and this falls to ~2wt% at room temperature. Nagashima[26] used X-ray diffraction to study the precipitation reaction in an Mg-12.5wt%Ag alloy. He identified two types of G.P. zones and proposed the precipitation sequence

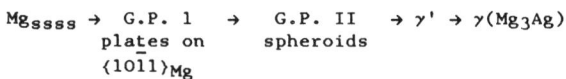

$$Mg_{ssss} \rightarrow \underset{\substack{\text{plates on} \\ \langle 10\bar{1}1 \rangle_{Mg}}}{\text{G.P. 1}} \rightarrow \underset{\text{spheroids}}{\text{G.P. II}} \rightarrow \gamma' \rightarrow \gamma(Mg_3Ag)$$

Gradwell[20] used transmission electron microscopy to examine the precipitation sequence in two Mg-Ag-Zr alloys, QK21 (Mg-2.41Ag-0.15Zr) and QK91 (Mg-8.73Ag-0.5Zr). In the concentrated alloy heterogeneously distributed γ' precipitates were observed during ageing at 200°C. No G.P. zones were detected, even when the ageing temperature was dropped to 75°C. Gradwell[20] proposed that the ageing sequence was

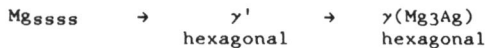

$$Mg_{ssss} \rightarrow \underset{\text{hexagonal}}{\gamma'} \rightarrow \underset{\text{hexagonal}}{\gamma(Mg_3Ag)}$$

Both Lagowski and Meier[27] and Gradwell[20] observed an approximately linear increase in U.T.S. with silver additions. This behaviour was attributed to solid solution strengthening, and almost no age hardening response. Certainly the coarse distribution of heterogeneously distributed γ' precipitates observed by Gradwell could not produce significant strengthening, Fig. 4.

The addition of 2wt%RE(Nd) to the Mg-Ag-Zr ternary produces an alloy with an outstanding ageing response and good tensile properties up to 200°C, e.g. QE22 (Mg-2.5wt%Ag-2.0wt%RE-0.6wt%Zr). The combination of silver and rare earth produces a high density of homogeneously distributed precipitates which are formed by a complicated precipitation sequence. Gradwell[20] investigated the distribution and crystal structure of the precipitates formed in QE22 at temperatures between 200 and 300°C. He identified two types of G.P. zones and two different intermediate precipitates, Fig. 5, and suggested that the precipitation sequence was

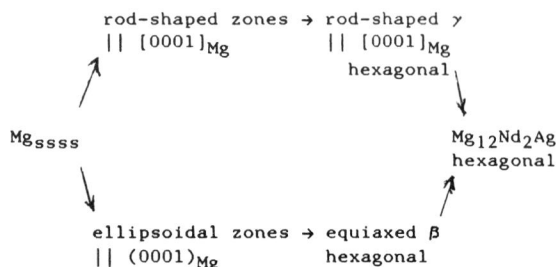

rod-shaped zones → rod-shaped γ
$|| [0001]_{Mg}$ $|| [0001]_{Mg}$
                hexagonal

Mg_{ssss} $Mg_{12}Nd_2Ag$
                           hexagonal

ellipsoidal zones → equiaxed β
$|| (0001)_{Mg}$ hexagonal

The two types of G.P. zone formed simultaneously during ageing at temperatures up to 250°C. Gradwell[20] proposed that the rod-shaped G.P. zones transformed into the rod-shaped γ intermediate precipitates, while the ellipsoidal G.P. zones transformed into the equiaxed β precipitates. No evidence was presented to illustrate the mechanism by which the zones transformed into the intermediate precipitates or how the transition from γ and β to $Mg_{12}Nd_2Ag$ occurred.

Gradwell attempted to determine the precipitation hardening mechanism in QE22 by examining foils taken from specimens which had been strained 2% in tension after ageing for various times at 200°C. He concluded that peak hardness coincided with the transition from precipitate cutting to Orowan looping.

CONCLUDING REMARKS

In most commercial aluminium alloys the precipitation reaction is well documented and the sequence of formation of successive precipitate phases is understood. This is not the situation in many cast, and heat treatable, magnesium alloys where there is still debate concerning the identification of the precipitate phases, e.g. in Mg-Ag, Mg-Th, Mg-Zn and their commercial derivatives. The details of the sequence of formation of precipitate phases, in particular the role of one precipitate in the nucleation of successive precipitates, has not been systematically investigated in most magnesium casting alloys.

Another development which is still in its infancy in magnesium casting alloys is the correlation of tensile and creep properties, including tensile and creep fracture, with microstructure. Magnesium casting alloys contain a eutectic at the grain boundaries, and remnants of the intermetallic component remain after heat treatment. These particles and both grain boundary and matrix precipitates formed during heat treatment will control ductility and fracture behaviour as well as tensile strength.

ACKNOWLEDGEMENTS

The author is indebted to Magnesium Elektron Ltd. for generous financial and material support for many years. Mr. W. Unsworth, Mr. J. King and the late Mr. S. Bradshaw, at the Swinton Laboratories of Magnesium Elektron Ltd. have been patient and knowledgeable collaborators. Drs. R. Worth (nee Moss), H. Karimzadeh and K. Gradwell have allowed the author to use illustrations from their Ph.D. theses.

REFERENCES

1. Emley, E.F., Principles of Magnesium Technology, Pergamon, London, 1966.
2. Polmear, I.J., Light Alloys Metallurgy of the Light Metals, Edward Arnold, London, 1981.
3. Stratford, D.J., in The Metallurgy of Light Alloys, The Institution of Metallurgists, London, 1983 p.132.
4. Waibel, J.H., in Metals Handbook, Desk Edition, ASM, Metals Park, Ohio, 1985, p. 8.1.

5. Clark, J.B., Acta Met. 16 1968 p.141.
6. Fox, F.A., J. Inst. Metals. 71 1945 p.415.
7. Sturkey, L. and Clark, J.B., J. Inst.
    Metals 88 1960 p.17.
8. Clark, J.B., Acta Met. 13 1965 p.1281.
9. Murakami, Y., Kawano, O. and Tamora, H.,
    Memoirs of Fac. of Eng. 23.93 Kyoto
    University, 1962.
10. Gallot, J., Ph.D. Thesis, Fac. of Sci.,
    University of Rouen, 1966.
11. Takahashi, T., Kojima, Y. and Takanishi, K.,
    Jap. J. Inst. Light Metals 23 1973 p.376.
12. Chun, J.S. and Byrne, J.G. Phil. Mag. 20
    1969 p.291.
13. Chun, J.S. and Byrne, J.G., J. Mat. Sci.,
    4 1969 p.861.
14. Mima, G. and Tanaka, Y., Trans. JIM
    12 1971 pp.71, 76 and 317.
15. W. Unsworth, This volume.
16. Moss, R.I., Ph.D. Thesis, University of
    Manchester, 1983.
17. Karimzadeh, H., Ph.D. Thesis, University
    of Manchester, 1985.

18. Mizer, D. and Clark, J.B., Trans. T.M
    A.I.M.E. 221 1961 p.607.
19. Mizer, D. and Peters, B., Met. Trans.
    3 1972 p.3262.
20. Gradwell, K.I., Ph.D. Thesis, Manchester
    University, 1972.
21. Pike, T.J. and Noble, B., J. Less Common
    Metals, 30 1973 p.63.
22. Karimzadeh, H., Worrall, J., Pilkington R.
    and Lorimer, G.W., This volume.
23. Kent, K. and Kelly, A., J. Inst. Metals,
    93 1964-65 p.540.
24. Mushovic, J., Ph.D., Thesis, Rensselaer
    Polytechnic Institute, Troy, N.Y. 1967.
25. Stratford, D.J., J. Inst. Metals, 100
    1972 p.381.
26. Nagashima, S., Trans.Jap.Inst.Met., 1, 1960
    p. 58.
27. Lagowski, B. and Meier, J., Canadian Dept.
    of Mines and Technical Surveys, Ottawa,
    R63, 1960.

Creep and high temperature properties of magnesium based alloys

B L MORDIKE and W HENNING

The authors are both in the Department of Materials Engineering and Technology, University of Clausthal, West Germany.

SYNOPSIS

The short and long term high temperature properties of magnesium base alloys are discussed. The discussion is limited to the most successful alloys to date. The influence of the various alloying elements on the high temperature strength and creep resistance is presented and the state of development of the magnesium rare earth alloys discussed.

INTRODUCTION

Magnesium alloys are used extensively in the aircraft industry and for space applications on account of the favourable strength:weight ratio. Good creep and high temperature strength are required of constructional components in applications such as transmission or gear boxes. Magnesium alloys melt, depending on the composition, between 743 and 923 K and this limits severely high temperature applications. Nevertheless, noteworthy achievements have been made as demonstrated, for example by the use of ZT1 (3% Th, 2% Zn, 0.7% Zr) for compressor housing in jet engines or the use of QE22 (2.5% Ag, 2% Nd rich rare earth, 0.7% Zr) for rotor heads of helicopters. Another high temperature application ($\sim$ 700 K) was realized with the start of nuclear power in Britain in the gas cooled nuclear reactors. Magnesium was chosen because of its low neutron cross sectional area and resistance to the fuel element material, uranium.

The containers were to be produced by extrusion and hence high strength and high ductility were required. This was achieved in the alloy Mg-0.8% Al, 0.005% Be (2). The addition of aluminium ensured sufficient strength due to solid solution hardening whereas Be was necessary to provide sufficient oxidation resistance. The melting point is hardly changed due to the small amount of alloying additions.

This is important as creep deformation becomes marked at operating temperatures in excess of 0.4-0.5 Tm. It should, however, be pointed out that the Magnox-nuclear fuel containers are subjected to very low static and dynamic loading compared with components in aerospace applications. If in addition to high temperatures high load carrying capacity is required then much more complex alloys such as the ZT1 alloy mentioned above must be developed. Fig. 1 shows the stress-time relationship for the 0.2% creep strain limit for ZT1 (3) at 573 K together with those for the cast aluminium alloy Al-4% Cu, 2.2% Ni, 1.5% Mn (4) and sintered aluminium powder SAP (5). It is clear that magnesium alloys behave comparably well with the normal high temperature aluminium alloys but not with the highly developed aluminium alloys.

The good creep resistance of Th containing materials is due to the Th phase which precipitates discontinuously on the grain boundaries. Ageing increases the strength further due to the precipitation of $Mg_{23}Th_6$ within the grains.

The precipitate with a melting point of 1045 K is one of the most thermally stable magnesium intermetallic compounds. Despite the grain refining effect of Zr the room temperature and elevated temperature tensile properties are poorer than those of other magnesium alloys, see Fig. 2. An optimisation of the short time properties is obtained by a Zn/Th ratio of 1.4:1 and this led to the development of the TZ6 alloy with 2.5% Zn, 1.8% Th and 0.7% Zr. This alloy, however, was less creep resistant than the ZT1.

Improved short time properties are also exhibited by QH21 (6) and QE22 (7). Both alloys are essentially further developments of ZT1 with the aim of combining good creep and high temperature tensile properties. In the case of QH21 (2.5% Ag, 1% Th, 1% Nd rich rare earth with 0.7% Zr) the zinc was replaced by silver and some of the thorium by Nd rich rare earth. In the case of QE22 with 2.5% Ag, 2% Nd and 0.7% Zr no thorium was retained. A solutionizing treatment followed by age hardening is necessary in both cases to develop optimum properties. The thermal stability of such precipitates, however, limits the creep resistance to a maximum of 523 K. Fig. 3 shows the creep of various magnesium alloys at 523 K. It is apparent that the thorium containing alloys are better than the Ag and Nd containing alloys. The alloy MTZ (1) is the zinc free variation of ZT1 and contains only 3% Th and 0.7% Zr.

The advantages of Th in relation to long term applications up to 623 K are obvious. Nevertheless, there are moves to replace the slightly radioactive Th by another non-radioactive element. The rare earth metals received most attention in recent years although the idea is not new. In the past, however, principally cerium misch metal (Ce, La and Pr) and didymium misch metal (with Pr and Nd) were considered (8).

Rare earth metals are candidates not only because of the chemical similarity and appropriate atomic size but also because they form high melting point intermetallic compounds with magnesium, which are similarly complex to those in the system Mg-Th. Table 1 (9) lists the most important intermetallic compounds. The silver containing compound Mg_3Ag has the lowest melting point of those compounds. $Mg_{23}Th_6$ is the thermally stable compound. These differences in the melting points are obviously related to the better creep behaviour of Th containing alloys.

The table also demonstrates that the properties of most rare earth metal systems can be improved by precipitation hardening. The first attempts were made with a casting alloy of the type Mg-Y-Zr obtained by simple substitution of Th in TZ6 by Y (10). As Fig. 4 shows the creep resistance could be increased by addition of Zn and Nd but did not surpass that of ZT1. Solution treatment of 24 h at 798 K with subsequent ageing for 24 h at 473 K produced marked improvement for these alloys (11).

In these age hardenable alloys zinc was eliminated as the precipitate required, $Mg_{24}Y_5$, did not form in the presence of zinc but rather the Mg-Y-Zn mixed compounds, eg. Mg_3YZn which are less thermally stable (12). The needle shaped, finely divided precipitate shown in Fig. 5 is a metastable form of the equilibrium precipitate $Mg_{24}Y_5$. These precipitates, formed by ageing at 473 K, produced the largest improvement in creep resistance (11).

Fig. 6 compares the yield and tensile strengths of YEK 741 (Mg-7% Y-4% Nd, 0.7% Zr) in the as-cast, homogenised and aged states over a range of temperatures (13). The aged specimens are clearly the best. Noteworthy is the maximum in the tensile strength at about 473 K which can be explained by an increase in ductility as a result of ageing or precipitation particularly in the case of homogenised or cast material.

A comparison of the yield stresses of YEK 741 of likewise aged QE22 and ZT1 demonstrates that so far as short time high temperature properties are concerned the Mg-Y-Nd-Zr alloys are clearly foremost (Fig. 7). The Zr free alloy YE62 (6% Y, 2% Nd) is also shown in Fig. 7 and is clearly weaker than YEK 741 but nevertheless markedly better than the Th containing alloys.

The Zr containing alloys exhibit, however, poorer creep properties. The reason for this is to be found in the strong coarse precipitate structure formed at 473 K due to the relatively high diffusivity of Y in Mg. The coarsening of the precipitates above ∿ 550 K leads to a strong temperature dependence of the stationary creep rate, as shown in Fig. 8. This restricts the high temperature application (14). The Zr free Y and Nd alloys are more successful in this respect. They are coarse grained and solidify dendritically. After solution treatment the grain size is medium large (300 μm).

Fig. 9 compares Zr containing alloys YEK 741 and ZT1 and the Zr free alloy YE 62 (Mg 6% Y-2% Nd). The good creep properties of the YE 62 can be attributed to the coarse grain. The age hardening cannot be responsible as coarsening has occurred at 573 K. An improvement in the creep resistance of ZT1 by omitting zirconium is not possible as the loss in room temperature properties would not be acceptable. Consequently YE 62 can be considered as an alternative material to ZT1.

Magnesium-rare earth alloys are being intensively studied in the USSR with emphasis on binary, ternary and complex systems containing Sm, Er, Tb and Gd. Only sparse information on the creep behaviour is available. It has been reported that alloys with up to 20% Gd or Tb can attain a 100 h rupture at stress of 120 N/μm at 523 K (9).

Specimens fabricated from QE 22 would fracture under these conditions in 25 min and those from MTZ in 25 h. Various authors have provided details of the tensile properties of these alloys at RT and 523 K (15,16,17). Fig. 10 compares the most successful alloys.

All alloys presented are in the aged or hot worked state. The short time properties are outstanding. Higher strengths were observed than for the Mg-Y-Nd-Zr alloys. Further experiments particularly on creep properties will show what can finally be achieved with Mg-RE alloys.

The future appears promising but where creep resistance is concerned the thermal stability is of particular importance. The diffusivity of the alloying elements is also important in determining the stability of the precipitates and hence directly and indirectly the creep rate. The high diffusivity of Y in Mg was a disadvantage in developing creep resistant Mg-Y-Nd-Zr alloys. The thermal stability exhibited by the $Mg_{23}Th_6$ precipitate may prove difficult to equal.

REFERENCES

/1/ E. F. Emley: Principles of the Magnesium Technology, Pergamon Press, Oxford (1966)

/2/ I. J. Polmear: Light Alloys, Metallurgy of the Light Metals, Edward Arnold Ltd., London, (1981), 156

/3/ MEL: Data sheet 456, Magnesium Elektron Ltd., Manchester, England (1982)

/4/ E. A. Brandes: Smithells Metals Reference Book, Butterworth and Co Ltd., London, (1983), 22-20

/5/ Aluminium-Zentrale Düsseldorf: Aluminium Taschenbuch, Aluminium-Verlag GmbH, Düsseldorf, (1974), 301

/6/ MEL: Data sheet 460, Magnesium Elektron Ltd., Manchester, England (1979)

/7/ MEL: Data sheet 464, Magnesium Elektron Ltd., Manchester, England (1982)

/8/ E. F. Emley, Gießerei, 23, (1969), 663

/9/ M. E. Drits, L. L. Rokhlin, A. A. Oreshkina, N. I. Nikitina, Russian Metallurgy, 5, (1982), 83

/10/ J. E. Morgan, B. L. Mordike, Metallurgical Transaction A, 12 A, (1981), 1581

/11/ B. L. Mordike, I. Stulikova, Proc. of the Int. Conf. on " The Metallurgy of Light Alloys, Loughtborought, England, (1983)

/12/ E. M. Padezhova, E. V. Mel`nik, R. A. Miliyevskiy, T. V. Dobatkina, V. V. Kinzhiballo, Russian Metallurgy, 4, (1982), 185

/13/ W. Henning: Untersuchungen zum Aufbau und zum Hochtemperaturverhalten von Magnesium-Seltenen Erden-Legierungen des Typs Mg-Y-Nd(-Zr), Dissertation, Technische Universität Clausthal, (Nov. 1986)

/14/ W. Henning, B. L. Mordike, Proc. of the Int. Conf. on " Strength of Metals and Alloys", Montreal, Canada, (1985), 803

/15/ L. L. Rokhlin, Russian Metallurgy, 4, (1979), 165

/16/ M. E. Drits, L. L. Rokhlin, E. M. Padezhanova, L. A. Guzei, Metal Science and Heat Treatment, 9, (1978), 771

/17/ M. E. Drits, Z. A. Sviderskaya, L. L. Rokhlin, N. I. Nikitina, Metal Science and Heat Treatment, 11, (1979), 887

Table 1: Binary Magnesium Alloy Systems

System	Max. Solubility (wt.-%)	Precipitation	Melting point of the Precipitation (K)
Mg-AG	15.5	Mg_3Ag	765
Mg-Ce	0.7	$Mg_{12}Ce$	884
Mg-Dy	25.8	$Mg_{24}Dy_5$	883
Mg-Er	32.7	$Mg_{24}Er_5$	857
Mg-Gd	23.5	Mg_6Gd	913
Mg-Ho	28.0	$Mg_{24}Ho_5$	838
Mg-Lu	41.0	$Mg_{24}Lu_5$	–
Mg-Nd	3.6	$Mg_{41}Nd_5$	833
Mg-Pr	1.7	$Mg_{12}Pr$	858
Mg-Sm	5.8	Mg_6Sm	835
Mg-Tb	24.0	$Mg_{24}Tb_5$	832
Mg-Th	5.0	$Mg_{23}Th_6$	1045
Mg-Tm	31.8	$Mg_{24}Tm_5$	865
Mg-Y	12.0	$Mg_{24}Y_5$	838
Mg-Yb	3.3	Mg_2Yb	991

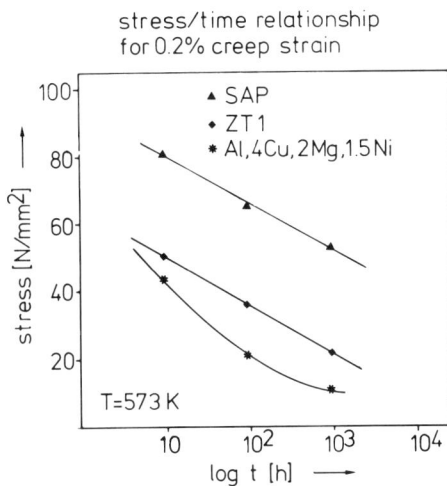

2 Comparison of the yield stresses as function of temperature for various high temperature magnesium alloys

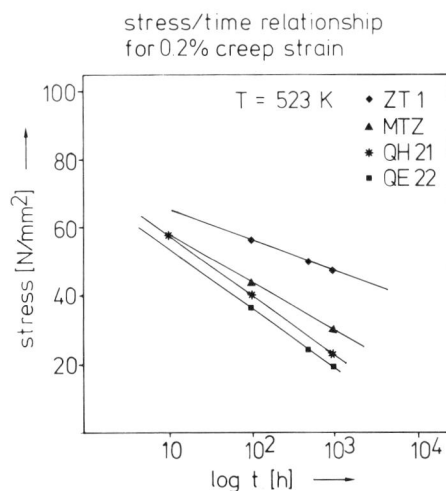

1 Time-stress relationship for 0.2 creep strain at 573 K for sintered aluminium powder (SAP), ZT1 and the high temperature alloy Al-4%Cu2%Mg1.5 %Ni

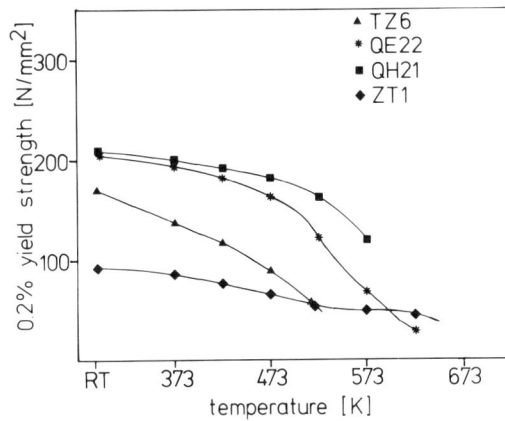

3 Time-stress relationship at 523 K for various high temperature magnesium alloys

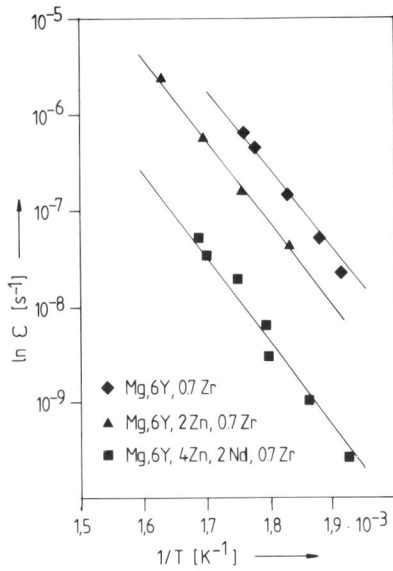

4 Comparison of stationary creep rates
over the temperature range 500 - 600 K
for various alloys based on Mg-Y-Zr
for an applied stress of 40 N/mm^2

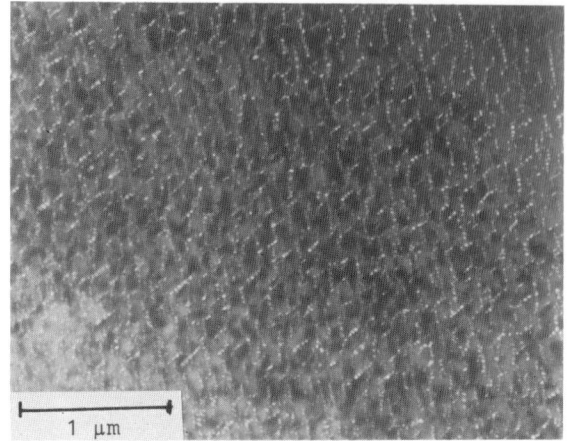

5 Needle shaped, metastable precipitates
of the equilibrium precipitate
$Mg_{24}Y_{5m}$ Mg6%Y4%Nd homogenised and aged

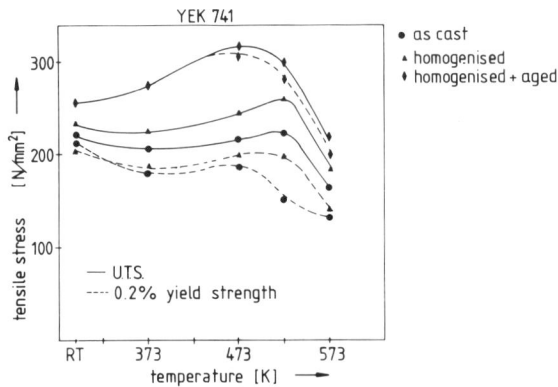

6 Yield stress and tensile strength as
function of the temperature for alloy
Mg7%Y,4%Nd,0.7%Zr (YEK 741)

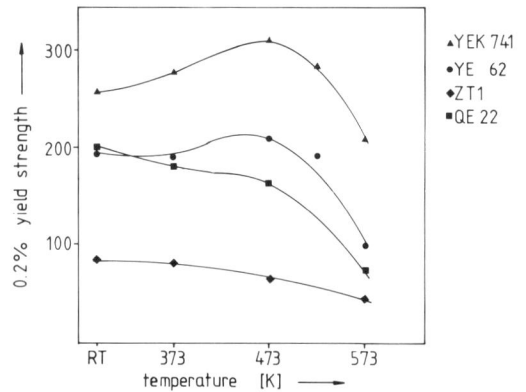

7 Comparison of yield stress as function
of the temperature for various magnesium alloys

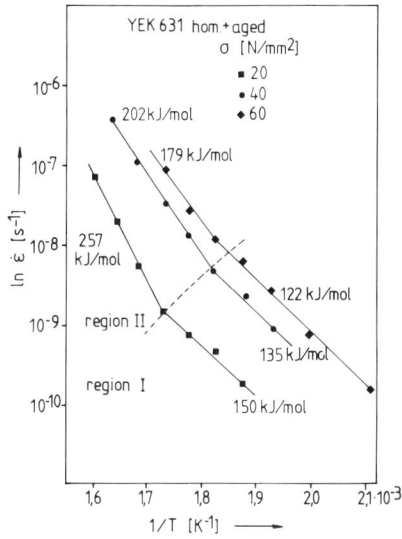

8 Stationary creep rates as function of
the temperature for various loads
Mg6%Y3%Nd0.7%Zr (YEK 631)

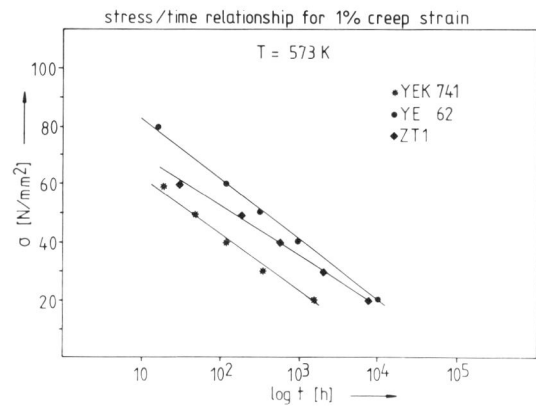

9 Time-stress relationship for 1 % creep
strain at 573 K for ZT1, YEK 741 and
YE 62

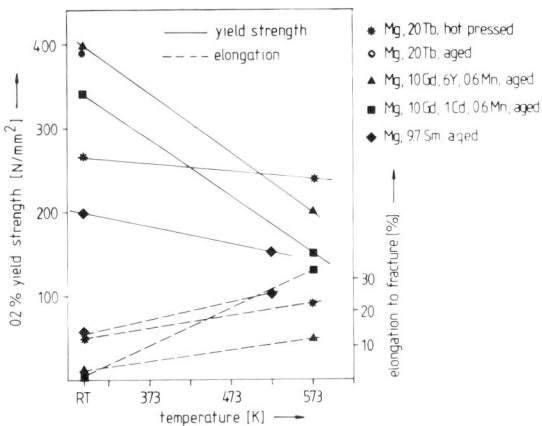

10 Yield stress and strain to fracture
as function of the temperature for
various magnesium base rare earth
alloys

Designing with magnesium alloys
G T HUDSON

*The author (retired) was formerly Executive
Director Technical, British Aerospace,
Kingston upon Thames, England.*

SYNOPSIS

Magnesium alloys in cast and wrought form have
been used in the manufacture of aircraft
components since the early days. Experience has
shown that such applications can achieve very
successful results giving both weight and cost
savings.

The use of castings, extrusions, or forgings has
the advantage of saving the more numerous parts
required in the fabrication of components made up
from sheet metal, plate and small machined parts
and subsequently rivetted or bolted together.
This usually results in reduction in costs and
often in weight.

In some applications, the minimum thicknesses
achievable in either casting, extruding, forging
or machining are limiting and in these cases the
lower density of magnesium alloys compared with
aluminium alloys gives a direct weight saving.
Again in the case of thin skins or webs subjected
to comprehensive or shear loadings in which
resistance to buckling is critical, the lower
density of magnesium alloys can, at moderate or
light loading intensities also save weight.
However great care needs to be exercised in the
choice of application in respect of the corrosion
susceptibility of magnesium alloys.

This presentation will I hope give some guidance
to designers on the way in which magnesium can be
used to advantage in applications where minimal
weight is an overriding criterion.

INTRODUCTION

The paper presents the results of my experience in
the design of missiles and military fixed wing
aircraft parts in magnesium alloys over a period
of some 35 years. It will include comments on
design and manufacturing and also the important
aspect of "In Service Experience".

The projects covered are :-

* An experimental supersonic missile, designed
  and tested over the period 1946 to 1950.

* The Folland Gnat Fighter and Trainer
  derivative, designed in the early 1950's and in
  service to the mid 1970s.

* The Hawker Siddeley Harrier and derivatives,
  initial design mid-1950's to mid-1960's, still
  under development and ongoing in service.

* The Hawk and derivatives, designed in the early
  1970's still under development and ongoing in
  service.

Experimental Missile (Designated RTV2)

The primary purpose of this supersonic missile was
to flight test a new design of rocket motor which
used kerosine and hydrogen peroxide (HTP) as a
propellant. The design was subcontracted to
Folland Aircraft at Hamble from the Royal Aircraft
Establishment. The vehicle was required to
achieve a definitive range and supersonic speeds.
It required a formidable battery of solid
propellant booster rockets in order to achieve a
controlled flight speed from a short launcher rail
and was required to manoeuvre at up to 15g
acceleration. The company were responsible for
the structural design, performance, the propellant
system, the integration into the design of the
rocket motor, booster rocket system and a guidance
and control system.

Preliminary project studies showed the need for an
HTP capacity of some 20 to 30 cu ft, the kerosine
capacity being small by comparison. The dictates
of stability and hence centre of gravity position
during propellant usage obviously required that
the HTP be contained over the midships region
coinciding with the location of the main lifting
wings.

The layout is shown in Figure No.1 (booster
rockets omitted for clarity). The dimensions were
body length about 18 ft and diameter 17 inches,
cruciform main lifting surfaces of approximate
span 6 ft and chord 4 ft.

The structural debate centred around the
containment and control of HTP. It was decided
that this was to be fed to the rocket motor by
means of pressurising nitrogen and that a smooth
internal bore was required in which a separator
piston would operate. The four wings with
substantial root bending moment were to be mounted
on this cylindrical HTP tank.

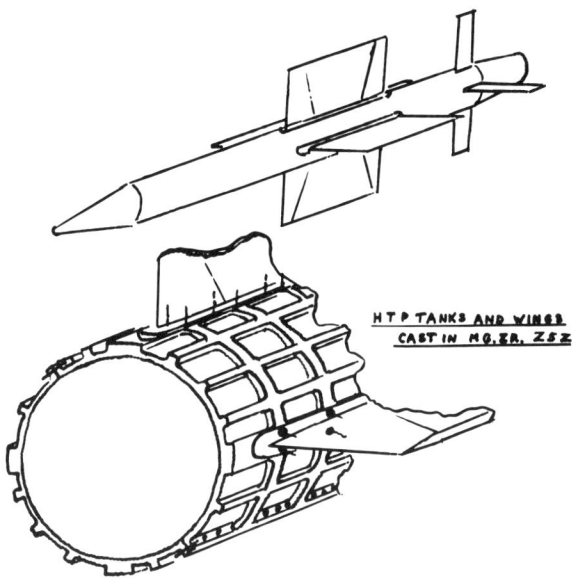

1.   Experimental Missile (RTV2)

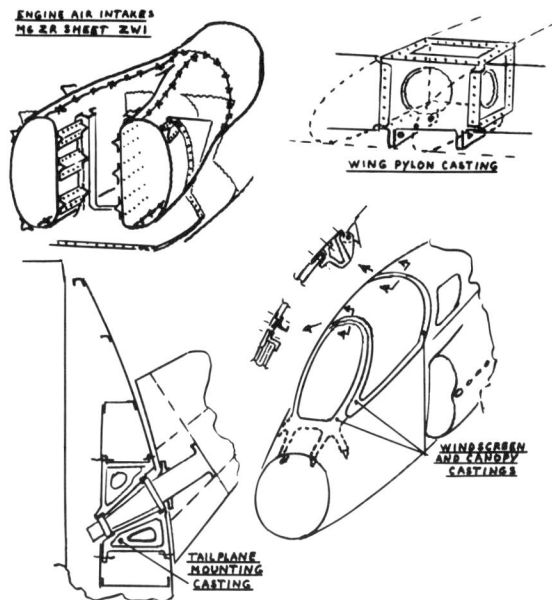

4.   Details of Gnat Magnesium Parts

2.   RTV2 missile recovered after firing

3.   Gnat Trainer RAF Red Arrows

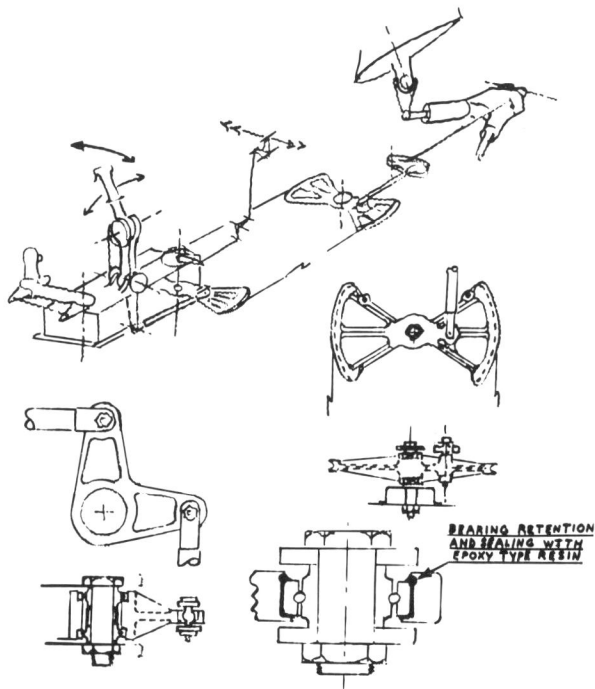

5.   Gnat Flying Controls

Extensive studies into alternative designs of both HTP tanks and wings covered fabrication with machined ring and inner and outer skins for the tank and similarly machined spars and ribs with sheet skins covers for the wings or the alternative, light alloy castings for both components. In the event the solid cast components won the day on two major counts; assembly difficulty and costs of fabrication and also the greatly increased thickness resulting from fabrication, leading to substantial drag penalties at high speed. A comparison between the then available cast aluminium and magnesium alloys showed a weight advantage for magnesium. Figure No.1 shows inset details of tank and wing surfaces and method of attachment.

The body forward and aft of the cast HTP tank was designed as a conventional sheet metal shell, reinforced with internal ring frames. In this case the geometry and loadings were such that compressive and shear buckling were critical and hence magnesium was selected due to its superior specific buckling resistance. It was also felt that the use of the same and compatible material throughout this integrated structure was a sensible choice.

The decision to use magnesium was accepted and encouraged by the Ministry on the basis that it was an indigenous material which could be extracted from seawater. Some 20 or 30 missiles were produced and many successfully fired.

The project demonstrated the benefits in manufacturing costs of substituting one piece component for conventional skin, stringers, frame, spar and rib assemblies with vastly increased numbers of detail parts and attendant assembly difficulties in essentially thin supersonic surfaces. The missiles did not require or demonstrate a long life in terms of exposure to corrosion, they were however simple and easily protected parts and would with suitable precautions have been satisfactory. Figure No.2 shows a missile recovered from the sea after firing.

Gnat Fighter & the derivative Gnat Trainer (Figure No.3)

W.E.W. (Teddy) Petter arrived at Follands in 1950 and his brainchild, the small lightweight Gnat Fighter took shape under his leadership. Petter had used magnesium alloys in his previous designs and was pleased to find design and manufacturing expertise existing at Hamble, together incidentally with a quantity of magnesium alloy sheet surplus from the RTV2 programs and therefore available at a keen price. A relatively large number of parts were designed in magnesium. Some of these are illustrated in Figure No.4.

The largest and perhaps the most ambitious component was the bifurcated engine air intake, from the entry on either side of the cockpit to the circular engine compressor face. The shape was almost entirely complex double curvature, as shown in Figure No.4 and magnesium zirconium sheet (ZW1) proved an excellent material in manufacture. The complex shapes were formed in large panels easily and accurately, using hot dies at, I recollect, 300 Deg. C or so. The formed panels were argon arc welded along the lines indicated. Alas the in service experience was not without its

troubles. In the case of the RAF Gnat Trainer version the combination of airborne sand and salt atmosphere at RAF Anglesey proved too severe. Corrosion reared its ugly head after a few years and the situation was only checked and recovered with a great deal of hard work by the ground maintenance teams. In retrospect, the combination of slightly dissimilar solid rivets and damage to the protective treatment by percussion rivetting together with the Valley environment combined to give trouble.

The only other Gnat magnesium component to cause significant difficulties due to corrosion was the built in wing pylon attachment fitting. It was not possible to provide adequate drainage and ventilation and moisture due to condensation eventually resulted in corrosion at the wing lower skin flanges.

Now for the good news. A large casting housing the rear fuselage tailplane spigot bearings was well drained and ventilated. It was fastened to the rear fuselage structure by bolts rather than rivets and gave totally satisfactory service. The cantilevered windscreen frame and canopy arch members also gave satisfactory service and showed substantial cost and weight savings compared with fabricated components of complex shape.

Finally in the Gnat a large number of components in the primary flying control system were produced as magnesium zirconium castings as shown in Figure No.5. These included the control columns, and its 'pedestal' mounting and most of the numerous idling levers and wire pulleys. They showed a direct weight saving over aluminium castings being produced at minimum casting thickness. These levers and pulleys required a minimum of machining for bearings and pulley wire grooves and were manufactured at a fraction of the cost of fabricated parts. One outstanding feature of these designs was a novel method of securing and sealing the outer races of the numerous ball bearings involved as shown typically in Figure No.5. This method of securing bearings gave two major advantages:-

a. The Mg - steel interface was totally sealed to preclude fretting due to relative movement and

   any consequent corrosion.

b. Very close tolerance housing bores were not necessary.

Extensive testing was carried out to establish the most suitable resin formulation and the optimum diametric clearance between bearing and housing. This involved environmental conditions including temperature cycling representing ground soak, high speed and high altitude flight giving temperature ranges from about - 60 Deg C to over 100 Deg C. The optimum diametric clearance was found to be about 0.003 to 0.004 inches per inch of bearing diameter. Replacement of bearings was found to be a straightforward operation using an automatic electrical heating plug through the bearing inner bore softening the resin sufficiently for removal at about 200 Deg C.

Note however that the design was such that in the event of bond failure the bearing could not disengage from the housing.

6. RAF Harrier GR mk 3

7. Sea Harrier

8. Hawk T Mk 1 RAF Red Arrows

9. Hawk Windscreen and Canopy

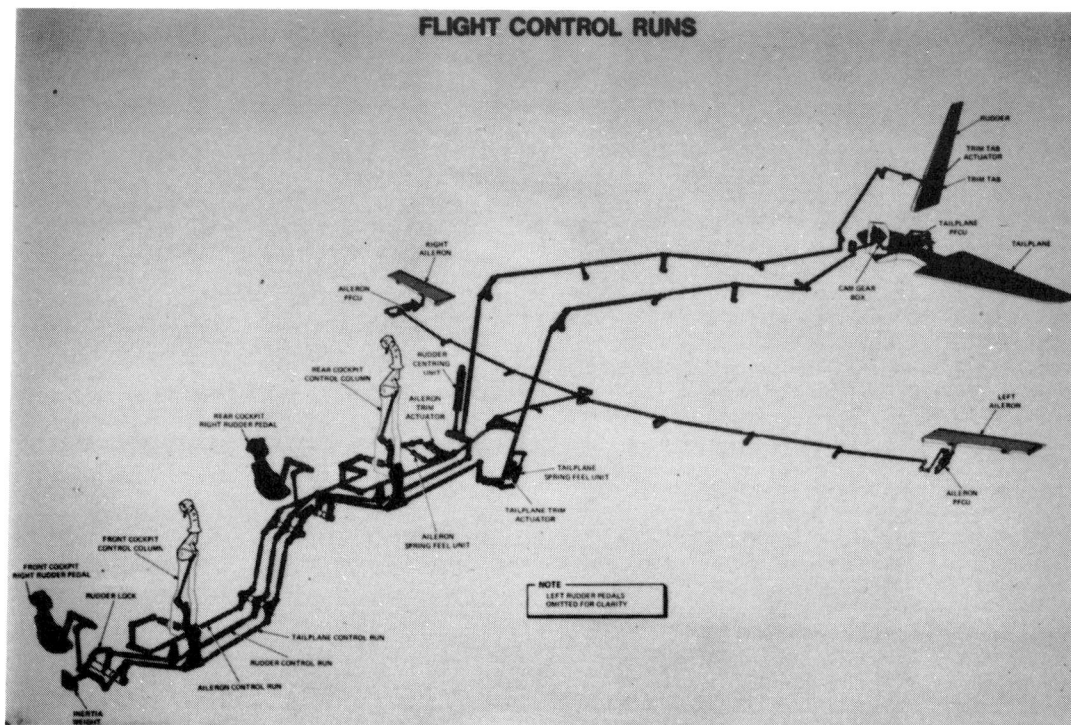

10. Hawk Flying Control System

Completely isolated from other structures and operating in well ventilated bays, these primary flying control components proved totally satisfactory in service and an ideal application for magnesium alloys.

Some 250 Gnat Fighters were produced and operated by the Indian Air Force and about 100 Trainer derivatives were in service with the R.A.F. for 15 to 20 years in both cases.

Harrier and Derivatives (GR mk 1 and 3. Sea Harrier, AV8B/GR mk 5) Figure No.6

There is not a great deal to say about this excellent and successful aircraft in relation to magnesium. That vastly experienced design team at Kingston upon Thames had many well tried and established practices. The use of magnesium did not feature in their philosophy. I believe from memory that the only significant item on this airframe is the cast magnesium alloy nosewheel. Satisfactory' in land based use but probably justifiably changed to aluminium alloy for the Sea Harrier and AV8B where the aircraft are ankle deep in sea-water on Carrier decks.

See Figure No.7.

The Pegasus 103 engine as fitted to land based Harriers (GR mk 3 and AV8A) had a compressor casing in cast mg. alloy. This also was changed to Al alloy for the Pegasus 104 for use in the Sea Harrier and AV8B.

Hawk and Derivatives (Figure No.8)

As Chief Designer of the Hawk, I was able to put past experience into practice. The Hawk therefore has magnesium alloy parts in applications where trouble free service was proven by Gnat experience. Thus magnesium castings are used for the windscreen side members and the canopy centre beam and rear arch, See Figure No.9.

The primary flying control system of the Hawk is signalled entirely by push rods (see Figure No.10) - this involves the use of a large number of idling and direction changing levers and based on Gnat experience we decided that magnesium alloy was the obvious choice.

It was however decided to produce these items as close tolerance enclosed die forgings rather than the sand castings used on the Gnat. This enclosed die technique was at the time well developed and offered two advantages compared with castings.

Firstly the superior strength of wrought material over castings and secondly, the closer control on item weight, achieved by weight control of the extruded billets prepared for forging. In this latter respect it is noted that in the enclosed die method there is no loss of material as in the case of split die forgings where material is extruded through the die lines in the form of a flash. The aircraft industry has been plagued for many years by flash line failures in die forged aluminium alloy forgings. Residual stresses due to heat treatment, assembly or continuous operating loads have led to stress cracking or stress corrosion cracking followed by fatigue in many primary structural parts. In these cases the very short grain found in the flash line region has been the main contributory cause. There have

also been cases of overheating in the die line, generated during forging which in some alloys can cause irrecoverable damage. Whilst we had no reason to believe that similar deficiencies could occur in magnesium alloy, it seemed a good idea at the time to avoid the possibility by the enclosed die technique.

Having said all this, I was extremely disappointed when it eventually transpired that the forgers could not economically obtain the extruded blanks. The minimum quantities on offer at the time would have produced several thousand Hawk aircraft sets. We therefore changed these parts to aluminium alloy at a cost in weight and the opportunity to use magnesium in this ideal application was lost. I understand that the lesson has now been learnt by the material suppliers and this unfortunate and hopefully temporary situation may now be overcome.

Relatively large magnesium castings were however used in the flying control system for the cockpit control pedestal components and most of the ball bearings in the system were assembled by the gluing in technique used in the Gnat system.

Some 350 Hawk aircraft have been produced to date and have been in service for up to 10 years in a number of Air Forces. With the selection of the Hawk for the new U.S. Navy trainer (T45) it is expected that total Hawk sales will approach 1000 over the next few years.

Concluding Remarks

The use of castings, extrusions or forgings is an effective way of saving the cost in manufacture and assembly of more numerous parts. Magnesium alloys can successfully be applied in many cases, saving weight compared with aluminium alloys. This is particularly the case where minimum attainable wall thicknesses in casting, forging or indeed machining are limiting rather than allowable stress levels. In thin panels and flanges, the superior specific buckling resistance of the lighter magnesium alloys will under moderate to light loading conditions also provide weight saving.

Do not specify magnesium alloys unless effective drainage and ventilation is obviously achievable and never in the bottom of asssemblies where water collects. These areas cannot effectively be drained dry however much effort is made.

Beware of percussion rivetting, it damages protective treatment and remember that protective coatings are not effective over sharp corners.

If minimum weight is of great importance, casting drawings should specify minimum allowable thicknesses together with maximum permissable raw casting weight. Foundries as well as machine shops display a tendency to err on the generous side in respect of thickness to avoid scrap. Other comments of significance in relation to castings concern quality and radiological examination.

It should always be borne in mind that some components have the highest tensile stress levels at the surface, notably parts subjected primarily to bending. In these cases surface defects are of much greater significance than blow holes or porosity deep within the casting.

On castings typical of those described in this paper, the cost of radiological examination may exceed the raw casting cost (or did 20 years ago). It is usually possible safely to reduce the extent of this examination after the casting technique has been shown to give a consistent product, thereafter examining on a sampling basis by agreement with the inspection or quality authority.

Finally should a magnesium alloy be developed in the future with improved corrosion resistance comparable with aluminium alloys then the don'ts emphasised in this paper would no longer be valid.

In this event, it would be worth considering research into "superplastic forming and diffusion bonding". This is now very successful in titanium alloys and under active research for aluminium. The advantage of the less dense magnesium in respect of buckling performance and its possible advantage in the more lightly loaded control surfaces against buffeting and acoustic fatigue may be of great importance.

Acknowledgements

My thanks are due to British Aerospace for the loan of some of the slides shown.

I acknowledge the excellent and hard work of many colleagues with whom I was associated on the projects described in this paper.

Corrosion and surface protection developments

D S TAWIL

The author is in the Metals Research Department at Magnesium Elektron Limited, Manchester, England.

SYNOPSIS

A review of developments in the area of corrosion and surface protection in the past 20 years is given. Alloy chemistry modifications to AZ91 and the development of the new high temperature alloy WE54 have produced materials with inherently good salt spray resistance. Surface chemistry modifications by Cd and Al ion plating and diffusion coating did not provide corrosion protection on Mg-Zr alloys but ion implantation of B and Fe into AZ91 has produced some promising results. The importance of surface cleaning prior to corrosion protection is emphasised. Chromate and anodic pretreatments are reviewed and some specialised coating systems for aerospace, automotive, computer and high temperature applications are highlighted.

INTRODUCTION

The perception of magnesium as a rapidly corroding material has been a major obstacle to its growth in structural applications despite its other obviously desirable physical properties. In fact, under normal environmental conditions, the corrosion resistance of magnesium alloys is comparable or better than that of mild steel (Table 1). It has been the uneducated use of magnesium in wet, salt laden environments which has given rise to its poor corrosion reputation[1]. Corrosion due to bad design, flux inclusions, surface contamination, galvanic couples and incorrectly applied or inadequate surface protection schemes are all avoidable occurrences, applicable not only to magnesium but to many other metals as well.

The magnesium industry has sought to educate the designer and engineer in the correct use of magnesium in corrosive environments[2,3,4] and has directed development work over the past 20 years to improve the natural corrosion resistance of magnesium alloys by modifications to alloy chemistry as well as improving surface protection schemes.

ALLOY CHEMISTRY

Typical corrosion rates for unprotected magnesium alloys as determined in accelerated tests (Table 2) show that magnesium is adversely affected in salt laden environments. While the correlation of these results to actual service performance is anything but well defined, the objective has always been to improve salt spray resistance.

The detrimental effects of heavy metal impurities and surface contamination on the corrosion performance of magnesium alloys have been known for some time[5] and speciality 'high purity' alloys with extremely good corrosion resistance were developed for use in the nuclear industry. Recently, this concept has been applied to the most commercially used magnesium alloy - AZ91 (Mg-9% Al, 1% Zn-Mn). The adverse effects of Fe, Ni and Cu in this alloy system have been quantitatively studied[6] and threshold levels established (Table 3) below which a 50 to 100 fold improvement in salt fog corrosion resistance is obtained.

The Fe threshold level was found to depend on the manganese content of the alloy. A sharp increase to high corrosion rates was observed when Fe levels exceeded 3.2% of the manganese content.

The copper threshold has been set at 300 ppm but it is known that higher levels can be tolerated if the zinc content is above the specification minimum of 0.4%.

The Ni threshold level was independent of the manganese or other major alloying addition but it was dependent on the casting form. The slower solidification rates for sand and permanent mould castings significantly reduced Ni tolerance to 10 ppm compared to 50 ppm for high pressure die castings.

In addition to the individual contaminant threshold levels, the corrosion rate was also affected by heat treatment. High purity AZ91 castings in the T6 and T5 condition had lower corrosion rates compared to castings in the as cast or T4 solution treated state.

The attainment and control of these low impurity levels in both high pressure die and sand castings has been demonstrated[7,8], resulting in the ASTM designation of the high purity alloys AZ91D and AZ91E respectively.

Pressure die cast components in AZ91D have been in service[9] for several months in automotive and computer applications. AZ91E sand castings have recently been specified for helicopter application. The high purity concept is being extended[10] to the die casting alloy AM60 (Mg-6% Al-Mn) and to other alloys in the Mg-Al system.

In the Mg-Zr system the new high strength, high temperature sand casting alloy WE54 (Mg-5% Y, 4% RE-Zr) has a similarly low salt spray corrosion rate. Fe, Si and to a lesser extent Ni impurities are naturally controlled[11] to very low levels in the presence of Zr. In the absence of Zn or other active alloying constituents the yttrium addition helps promote a protective oxide film.

The salt spray corrosion resistance of WE54 and the high purity AZ91 alloys compare favourably with those of aluminium alloys (Fig. 1). This considerable improvement does not however protect against galvanic corrosion. In corrosive environments the standard techniques to reduce or eliminate the effects of galvanic couples must still be employed[4,12]. The benefits of these "corrosion resistant" alloys will only be fully realised if existing surface protection levels are maintained for a given application. Under these circumstances corrosion resistant alloys will significantly limit corrosion spread from a damage point (Fig. 2) so that increased component life with reduced maintenance costs can be expected.

Perhaps rightfully considered in this section is the impact that fluxless melting has made on the overall corrosion resistance of magnesium alloys. Flux inclusions in castings, due to poor metal handling techniques have contributed to magnesium's bad corrosion reputation in the past. This has not been a problem with premium quality castings but has affected the service performance particularly of die cast material. The use of sulphurhexafluoride (SF_6) in a protective gas cover during melting and casting operations has significantly improved the metal quality achieved in pressure die castings.

The techniques of ion implantation, ion plating and diffusion coatings have also been used to modify alloy composition at the surface with the aim of improving corrosion resistance particularly at dissimilar metal contacts.

Ion plated coatings of cadmium and aluminium on ZE41 (Mg-4% Zn, 1% RE-Zr) and QE22 (Mg-2% Ag, 2% RE-Zr) alloys have been investigated. Preliminary experiments produced reasonably dense coatings initially with the tendency to become porous as film thickness increased. Corrosion evaluations quickly demonstrated the unsuitability of these metallic coatings which caused galvanic attack of the magnesium at damage points.

Attempts to diffuse these coatings into the substrate by subsequent heat treatment under inert atmosphere were only partially successful (Fig. 3) due to the presence of an oxide barrier at the substrate/coating interface. Argon glow discharge cleaning (off sputtering) parameters were established whereby the oxide barrier on the magnesium substrate was substantially removed. This cleaning treatment however, caused preferential etching of the zinc rich β phase

(Fig. 4) on ZE41 which prevented uniformly dense ion plated coatings being subsequently deposited. This effect did not occur with QE22 alloy after similar glow discharge cleaning.

The effective cleaning of the substrate enabled uniform diffusion of cadmium and aluminium ion plated coatings after heat treatment at temperatures of 225o and 400oC (Fig. 5) respectively. These diffusion coatings, with or without additional organic finish, were also unsuccessful in providing corrosion protection under salt spray conditions. Microprobe (SEM) analysis demonstrated that diffusion of the coating material into the matrix was accompanied by outwards diffusion of magnesium which prevented the formation of a gradual protective compositional gradient.

Ion implantation of iron[13] and boron[14] into AZ91 has recently been reported. Implantation of boron produced a significant reduction in current density at passivation but the surface began to pit at approximately the same voltage as the unimplanted samples. Surprisingly, implantation of iron was found to improve the corrosion resistance of AZ91C alloy as determined by anodic polarisation measurements in a dilute chloride containing solution. The current limitations of ion implantation techniques preclude its use for commercial application but the data yielded by such work may be of benefit for future modifications to alloy chemistry.

CLEANING TREATMENTS

The effects of heavy metal contamination resulting from shot, grit or sand blasting operations at the foundry can be particularly damaging to the corrosion resistance of magnesium alloys[15]. While there have been no major developments in surface cleaning treatments the significance of a cathodically clean surface is now even more important when using the new corrosion resistant alloys.

The established techniques of acid pickling and Fluoride Anodising[4] remain the standard methods for contaminant removal from the surface. Recent work[8] has demonstrated that as acid levels in pickling baths deplete, the tolerance to dissolved iron in the bath is reduced and its cleaning ability is impaired (Fig. 6). For example, a freshly prepared nitric-sulphuric acid pickling bath has an iron tolerance of approximately 250 ppm whereas after 50% acid depletion this tolerance is reduced to just 50 ppm. To avoid residual iron contamination after acid pickling a second stage immersion in 10% hydrofluoric acid, or better still the Fluoride Anodising of the component is recommended.

Alumina grit blasting is also used to clean and prepare magnesium surfaces. Providing high purity alumina is used, and the purity maintained, the risk of heavy metal contamination is avoided. In paint refurbishment programmes plastic bead or vegetable abrasive media (crushed nut shells) can be safely used on magnesium.

PRETREATMENTS

Conversion coatings are applied to magnesium in order to passivate the surface and provide a better 'key' for subsequent painting. Whilst conversion coatings will delay the onset of natural surface oxidation they should not be regarded as protective treatments in their own right.

Chromate conversion coatings, applied by simple dip technique, remain the most widely used pretreatment type. With the increased emphasis on the hazards associated with hexavalent chromium, chromate free pretreatments have been examined but with little success.

Some proprietary phosphate treatments are available[2,3] and have been commercially used on magnesium[9] but they are not as effective as chromating and consequently should be restricted for use in mild environments only.

The NH35[16] chromate treatment with reduced hexavalent chromium addition was developed for use on pressure die castings. This bath contains only 1/5th of the chromate traditionally used in similar treatments (Table 4) and should therefore help reduce chromate effluent arising from 'drag out' losses. Chromite coatings using trivalent chromium have recently been patented[17].

Of the many[18] hard anodising treatments developed for magnesium the HAE and Dow 17 remain the most commonly used. Both can be applied as thin or thick films. Thin films offer little advantage, if any, over conventional chromate conversion coatings but the thick films offer excellent abrasion resistance and the potential for improved corrosion resistance if used correctly.

The more recently developed MGZ[19] anodising process offers some advantages in treatment time and wear characteristics compared to the Dow 17 and HAE treatments (Fig. 7). Whilst the HAE film, is hard (750 VPN) and very abrasion resistant it is a rough film which can abrade materials brought into contact with it. In modified Taber abrasion evaluations with rubber wheels, the MGZ anodic film suffered much less wear and caused less rubber wear compared to the HAE film making it particularly suited for automotive pulley wheel applications.

COATINGS

It is beyond the scope of this paper to adequately cover the full range of conventional coatings and application methods now used to protect magnesium components in service. There are however some specialised coating systems which have given good service or are currently being developed for future magnesium applications.

The development of cathodic epoxy electrophoretic primers with their improved chemical and corrosion resistant properties compared to the earlier anodic systems have proved effective on magnesium[20].

Cathodic primers can be satisfactorily deposited onto chromate pretreated magnesium at thicknesses of 15-20 μm and will provide very good salt spray resistance. They have been used to protect magnesium components in the automotive industry[9] where the use of electroprimers is wide spread.

Tough, hard wearing nylon coatings were successfully adapted[21] for use on magnesium in aerospace applications. Complete long term protection of magnesium control pulley wheels, even in the aggressive environment of aircraft wheel bays, was achieved by applying the nylon coating over a fully leachable chromate inhibited epoxy primer system. The nylon can be applied by fluidised bed or electrostatic powder coating techniques to provide a very damage resistant barrier which allows magnesium components to be used under constant handling and rubbing conditions.

Wrought, high pressure die and investment cast magnesium components are used in computer applications where lightweight, low inertia, rigidity and heat sink requirements may preclude the use of other metals and plastics. Within disc drive units protective coatings must be applied to prevent even minute surface oxide particles from causing disc or head failure. Parylene[22] (polypara-xylylene) conformal thermoplastic coatings, originally developed for the electronics industry, have been effectively used to protect magnesium in these critical areas. The controlled, thin, uniform coatings are applied under partial vacuum conditions onto chromate pretreated substrates. They possess excellent gas and water vapour resistant characteristics and coatings as thin as 3 μm are sufficient for long term protection against atmospheric oxidation.

The technique of epoxy resin sealing[23] chromate conversion coatings, as a foundation treatment for the protection of magnesium components in aggressive environments, has been adapted for use with the hard anodic pretreatments.

Anodic coatings are porous and conventionally applied chromated primers (DTD 5567, MIL-P-23377) will not seal the porosity (Fig. 8). In the event of damage to the paint coating lying on top of the anodic film, corrodant could be absorbed into the porosity giving rise to subsurface lateral corrosion spread.

To prevent this happening the porosity in the anodic film must first be sealed. This is best achieved using thin, clear, high temperature curing resin systems applied by spray or dip techniques onto warm, prebaked anodised substrates. Although epoxy systems are preferred other resins can be used. Sealing the anodic film produces an extremely effective corrosion resistant duplex coating capable of providing many hundreds of hours salt spray resistance. Corrosion creepage from damage points is minimised and resin impregnation also toughens the hard but brittle anodic film further improving its damage and abrasion resistance.

Resin sealed anodic coatings have been used to protect magnesium components operating in very aggressive environments. One example (Fig. 9) is an atmospheric pressure deep sea diving suit whose body and helmet are in magnesium alloy. Surface protection consists of thick HAE anodic treatment, sealed with a thin clear high temperature baking epoxy resin, followed by chromated epoxy primer and epoxy top coat. Full wet assembly procedures are employed at dissimilar metal contacts. This protection scheme, coupled with good maintenance procedures, gives satisfactory service between major overhaul intervals every four years.

Impregnation of porous anodic films with MoS_2 or PTFE, as in the range of magnadize[24] coatings provides for hard wearing yet lubricative coatings with combined corrosion resistance.

One protection scheme, which is still under development, is for use at elevated temperatures up to 300°C. The scheme is based on Dow 17 anodic pretreatment sealed with a clear polyimide resin. Exceptional protection against hot aeroengine lubricants has also been indicated.

Although current aeroengine gearbox applications typically operate at temperatures up to 130°C future trends are for hotter gearboxes. The limiting factor for magnesium in these applications has not been lack of elevated temperature properties such as yield, fatigue and creep resistance but rather the problem of internal corrosive attack from the fatty acid breakdown products of ester lubricants.

Preliminary investigations showed that resin sealed chromate or HAE pretreatments were more adversely affected in hot oil environments compared to sealed Dow 17 coatings. A series of different resin systems on thick Dow 17 (Table 5) anodic pretreatment were evaluated in static lubricant immersion tests at elevated temperatures (Table 6). Corrosion of the test pieces was monitored by magnesium analysis of the oil.

At 170°C all the schemes based on Dow 17 completely withstood the oil test environment whereas the conventional epoxy sealer on chromate pretreatment exhibited complete breakdown. At 200°C all systems based on Dow 17 again gave very good performance but the Dow 17 + polyimide scheme was the only one to completely withstand the full 300 hour exposure. At 225°C all systems exhibited some breakdown but the Dow 17 + polyimide gave the best performance followed by the epoxy and phenolic sealed Dow 17 coatings (Fig. 10).

Further work under dynamic oil impingement testing at 250°C for 50 hours confirmed the polyimide sealed Dow 17 coating as the best scheme evaluated even with scribed damage to the coating. Environmental high temperature and salt spray corrosion evaluations showed that Dow 17, sealed with a clear polyimide resin, with an aluminium filled polyimide top coat was an excellent corrosion protection scheme for use up to 300°C. While further development of the polyimide system is required the advantages of Dow 17 pretreatment coupled with epoxy or phenolic sealers are immediately available.

FUTURE DEVELOPMENTS

The impact already made by the high purity alloy AZ91D in the North American automobile market and the great interest shown in AZ91E and WE54 alloys for sand casting application will ensure the development of other magnesium alloys with inherently good corrosion resistance. Rapid solidification techniques promise much in the field of improved corrosion resistance and a number of alloy systems are currently being examined. Continuing developments in resin and paint technology are expected to provide improved coating systems perhaps coupled with more effective corrosion inhibiting agents for magnesium.

In conclusion, the corrosion threat, be it real or perceived, will continue to be actively fought by the magnesium industry so that the full potential of magnesium's unique combination of properties can be utilised.

REFERENCES

(1) The Realities of Magnesium Corrosion and Protection - W. Unsworth and J.F. King, 41st Annual World Magnesium Congress, London, June 1984.

(2) Corrosion and Protection of Magnesium - AMAX Magnesium publication, 1984.

(3) Magnesium: Designing Around Corrosion - Dow Chemical Co. publication, 1982.

(4) Surface Treatments for Magnesium Alloys in Aerospace and Defence - Magnesium Elektron publication, 1983.

(5) Corrosion Studies of Magnesium and its Alloys - J.D. Hanawalt, C.E. Nelson, J.A. Peloubet. Trans Am. Inst. Mining Met. Eng, 147, 273-99 (1942).

(6) The Effects of Heavy Metal Contamination on Magnesium Corrosion Performance - J.E. Hillis. SAE Tech Paper 83-0523.

(7) Controlling the Salt Water Corrosion Performance of Magnesium AZ91 Alloy in High and Low Pressure Cast Form - J.E. Hillis, K.N. Reichek, K.J. Clark. SAE Tech Paper 85-0417.

(8) AZ91E Magnesium Sand Casting Alloy. The Standard for Excellent Corrosion Performance - K.J. Clark, Wellman Dynamics, Proc. I.M.A. Conference, Los Angeles, June 1986.

(9) Product Design and Development for Magnesium Die Castings - Dow Chemical Co. Publication.

(10) The Potential for Magnesium Die Castings in Aerospace Applications - D.M. Magers, AMAX Proc. IMA Conference, Los Angeles, June 1986.

(11) Principles of Magnesium Technology - E.F. Emley, Pergammon Press, 1966 p 176-190, 685.

(12) Metals Handbook - American Society for Metals 9th Ed. Vol. 2, 1979 Chapter IV P596-609.

(13) Effect of Fe Implantation on the Aqueous Corrosion of Magnesium - Akavipat, Hale, Haberman and Hagans, Mat. Sci. Eng. 69 (1985) p 311-316.

(14) Surface Modification of Magnesium for Corrosion Protection - P.L. Hagans, 41st Annual World Magnesium Congress, London, June 1984.

(15) Principles of Magnesium Technology - E.F. Emley, Pergammon Press 1966, p 692-696.

(16) The NH35 Chromating of Magnesium Pressure Die Castings - Norsk Hydro Publication.

(17) USP 4,569,699 - 1986.

(18) Principles of Magnesium Technology - E.F. Emley, Pergammon Press 1966, p 700.

(19) USP 3,791,942 - 1974.

(20) New Process to Protect Magnesium Against Corrosion - R. Hugot, SAE Tech. Paper 83-0598.

(21) Reflections and Projections (Thoughts on the use of Magnesium alloys in Airframes) - G.B. Evans, B.Ae. Proc. IMA Conference, Los Angeles 1986.

(22) Parylene - Product of Union Carbide Corp. applied by licensed finishers.

(23) Specification DTD 935, 'Surface Sealing of Magnesium Rich Alloys' - H.M.S.O.

(24) General Magnaplate Corp., Linden, N.J. 07036 U.S.A.

TABLE 1

RESULTS OF 2 1/2 YEAR ATMOSPHERIC EXPOSURES
(Data from Metals Handbook)

MATERIAL	CORROSION RATE (μm/year)		
	RURAL	MARINE	INDUSTRIAL
Low Carbon Steel	15	150	25
Magnesium (AZ31)	13	18	28
Aluminium (2024)	0.1	2	2

TABLE 2

TYPICAL CORROSION RATE DATA FOR MAGNESIUM ALLOYS IN MILD TO SEVERELY CORROSIVE ENVIRONMENTS

ENVIRONMENT	CORROSION RATE ($mg/cm^2/day$)
ASTM B117 Salt Fog	2-10
3% NaCl Immersion	1- 5
Intermittent Seawater Spray	0.1-0.2
Industrial Atmospheric	0.01-0.02

TABLE 3

HEAVY METAL THRESHOLD LEVELS FOR HIGH PURITY AZ91 CASTINGS

APPLICATION	MAX CONTAMINANT LEVEL (%)		
	Cu	Ni	Fe
Pressure Die Cast (AZ91D)	0.03	0.005	0.032 X Mn %
Sand, Permanent Mould Cast (AZ91E)	0.03	0.001	0.032 X Mn%

Mn - 0.17% min, 0.35% max

TABLE 4

CHROMATE TREATMENTS FOR MAGNESIUM PRESSURE DIE CASTINGS

CONSTITUENT (gm/litre)	CHROMATE TREATMENT		
	DOW 1	DOW 20	NH35
Sodium Bifluoride ($NaHF_2$)	-	15	2.5
Sodium Dichromate ($Na_2Cr_2O_7.2H_2O$)	180	180	35
Magnesium Sulphate ($MgSO_4.7H_2O$)	-	-	3
Aluminium Sulphate ($Al_2(SO_4)_3.14H_2O$)	-	10	-
Nitric Acid 70% HNO_3 S.G. 1.42	187 (cm^3)	125 (cm^3)	30 (cm^3)

TABLE 5

PROTECTION SCHEMES EVALUATED FOR ESTER LUBRICANT RESISTANCE

SCHEME	PRETREATMENT	RESIN SEALANT
1	Cr/Mn chromate	Chromated epoxy
2	Dow 17 anodic	None
3	Dow 17 anodic	Clear epoxy
4	Dow 17 anodic	Clear phenolic
5	Dow 17 anodic	Clear polyimide
6	Dow 17 anodic	Chromated epoxy silicone

TABLE 6 - OIL IMMERSION TEST PROGRAMME

TEST	Coated EZ33 alloy specimens immersed in hot oil contained in a sealed system.
OILS	DERD 2497 and MIL-L-23699C specification.
TEMPERATURE	170°, 200°, and 225°C.
TIME	300 hours with oil changes every 100 hours.
MONITORING	Oil - magnesium content and acidity. Specimens - weight change and visual appearance.

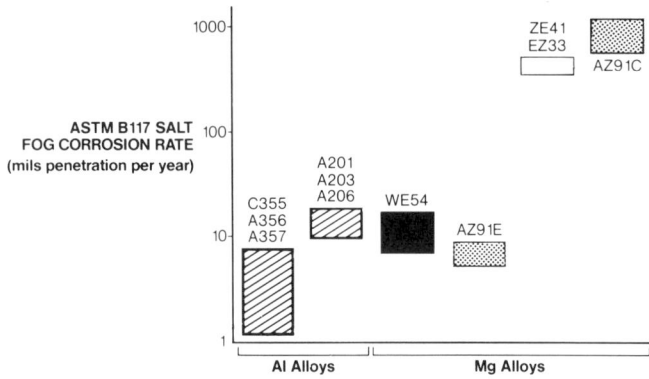

FIGURE 1 – CORROSION COMPARISON OF SOME
MAGNESIUM AND ALUMINIUM ALLOYS IN ASTM B117
SALT FOG

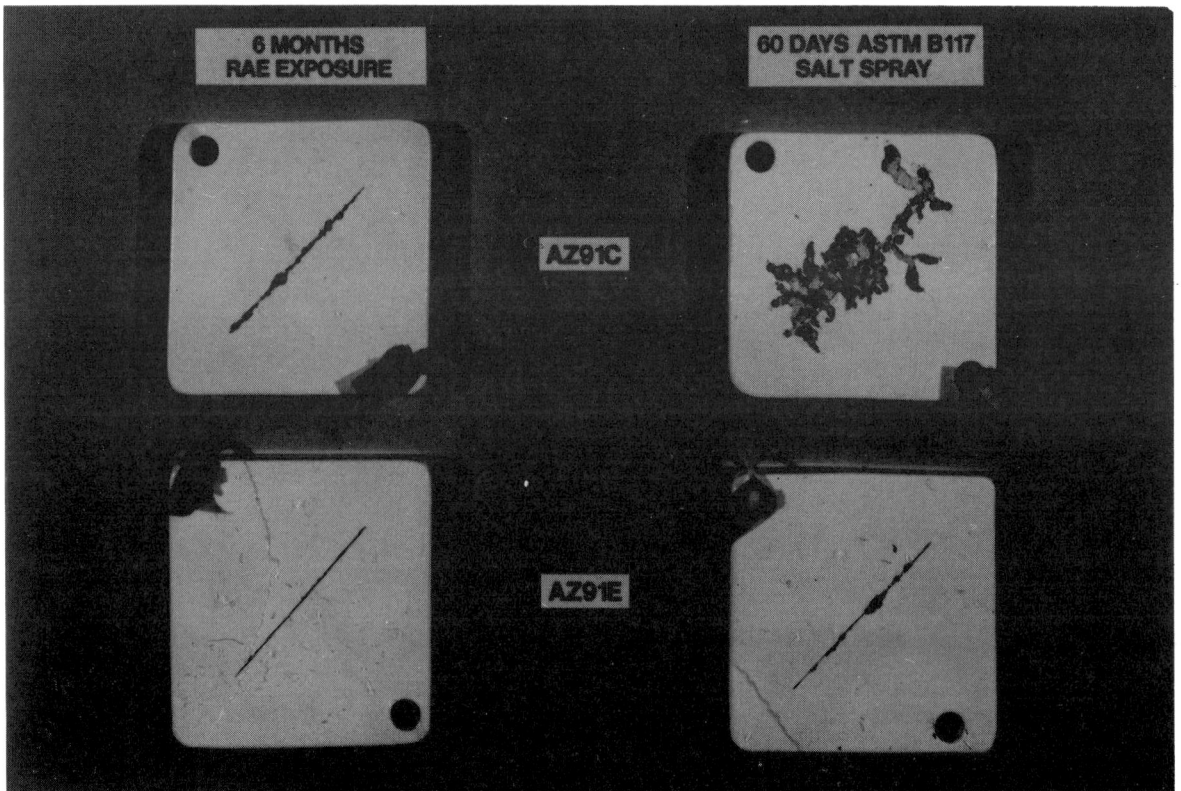

FIGURE 2 – PERFORMANCE COMPARISON OF AZ91C AND
AZ91E TEST PANELS PROTECTED TO DTD 911C
SPECIFICATION

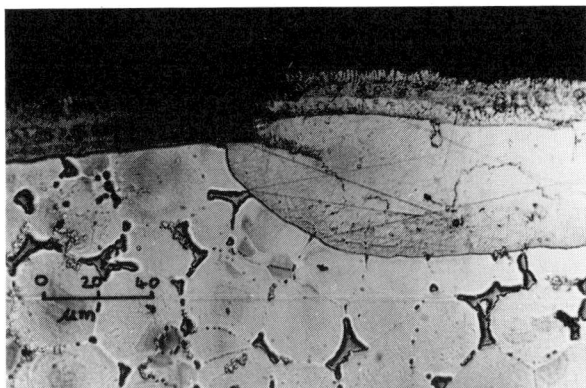

FIGURE 3 – INCOMPLETE DIFFUSION OF Cd COATING
INTO ZE41 DUE TO OXIDE BARRIER AT INTERFACE

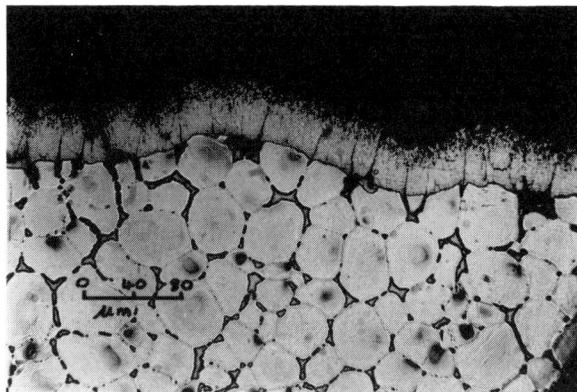

FIGURE 4 – PREFERENTIAL ETCHING OF Zn RICH β
PHASE IN ZE41 DUE TO ARGON GLOW DISCHARGE
CLEANING

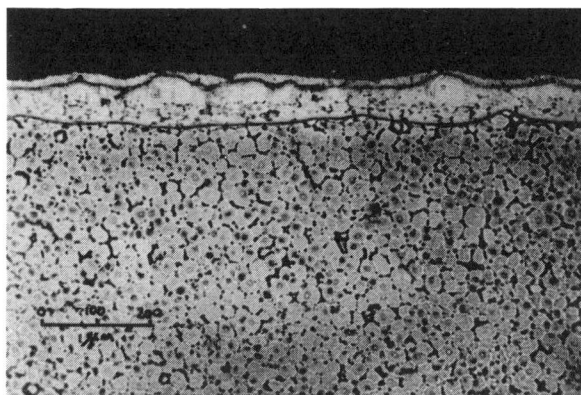

FIGURE 5 – UNIFORM 100 μ Cd DIFFUSION ZONE ON
ZE41

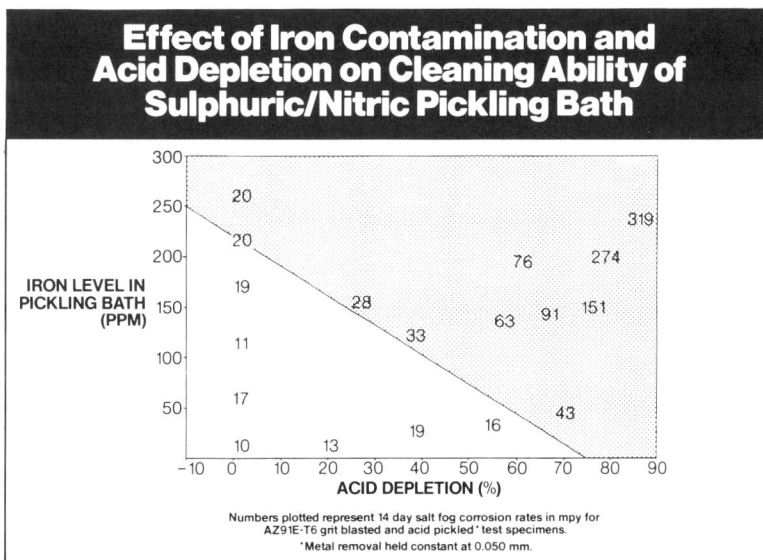

FIGURE 6

73

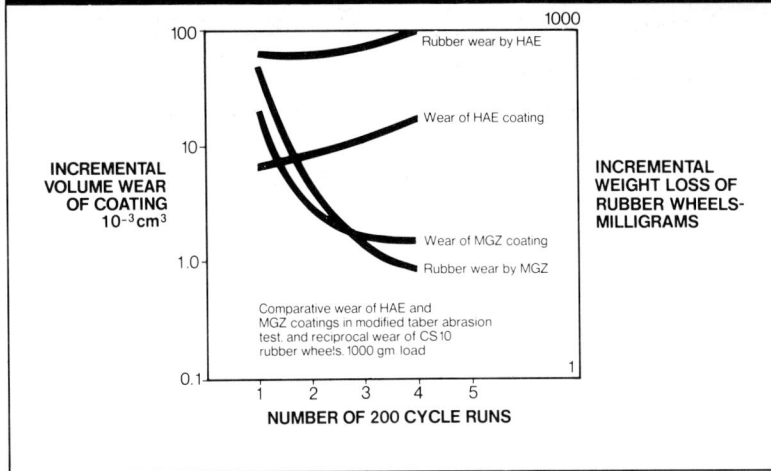

Comparison of Coating Wear and
Rubber Wear for MGZ and HAE Coatings

FIGURE 7

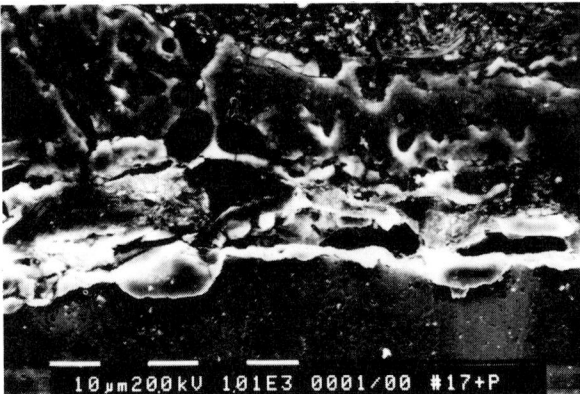

FIGURE 8 - FAILURE OF CHROMATED EPOXY PRIMER
(DTD5567, MIL-P-23377) TO SEAL POROSITY IN
DOW 17 ANODIC FILM

FIGURE 9 - EPOXY SEALED HAE ANODIC COATING USED
TO PROTECT MAGNESIUM BODIED DEEP SEA DIVING SUIT

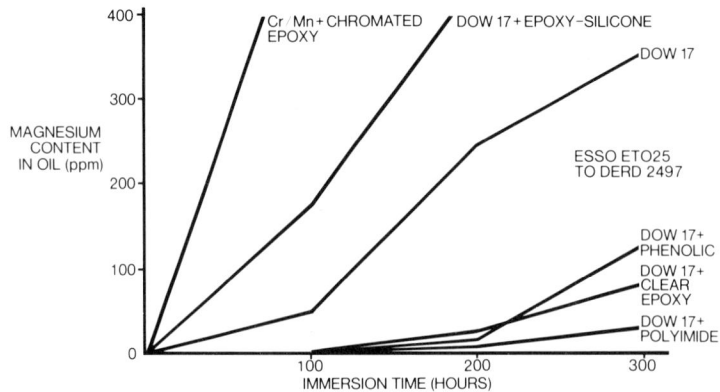

FIGURE 10 - RESULTS OF STATIC LUBRICANT
IMMERSION TESTS AT 225°C

Squeeze casting of magnesium alloys and magnesium based metal matrix composites

G A CHADWICK

The author is with Hi-Tec Metals R & D Limited, Chilworth Research Centre, Southampton, England.

SYNOPSIS

Several different magnesium alloys, both normal casting alloys and ostensibly wrought alloys, have been squeeze cast over a range of casting conditions and the mechanical properties of these cast materials have been determined. A new magnesium alloy specifically designed for the squeeze casting process has been produced and tested and has been shown to possess excellent creep and high temperature fatigue properties. Metal matrix composites based upon AZ91 and having graphite, glass and alumina as the reinforcing phase have also been produced and tested. The mechanical properties of these alloys in relation to their various microstructures are presented in this paper.

INTRODUCTION

Squeeze casting is the name given to that process in which metal is solidified under the direct action of pressure sufficient to prevent the appearance of either gas porosity or shrinkage porosity. Squeeze casting yields 100%-dense castings and the process is unique from that point of view.

The fundamentals of squeeze casting have been known for decades and the process has been used in Russia for 50 years or more. However, it is only recently that squeeze casting has been commercialised in the West and although many different metals and alloys have been tested experimentally its practical application has not yet received broad acceptance. The present paper deals exclusively with the squeeze casting of magnesium alloys and magnesium matrix composites.

Squeeze casting takes place in metal dies sufficiently thick to withstand the applied pressures: in the work reported here pressures from 0.1 MPa up to 300 MPa have been used. The dies are usually heated and lubricated, the lubrication medium commonly being a graphitic die coat. Figure 1 shows diagrammatically the squeeze casting process used to produce both solid ingots and hollow shapes. A metered amount of metal is poured into the lower die and the top die, or punch, is forced down onto the liquid metal, pressure being maintained throughout the freezing range. The casting is removed from the die either by using a fixture on the punch or by use of an ejector. With correct metering there is no metal wastage, since risers and runners are not used. Because of the high pressures involved, the metal casting replicates precisely all features of the die cavity and by the same token the heat transfer coefficient is higher than in other casting processes.

The high pressures used cause temperature and compositional changes to the phase transformations as obtained at atmospheric pressure. The change in freezing temperature is given by the Clausius-Clapeyron equation $\frac{dT}{dP} = \frac{T_f \Delta V}{H_f}$, where T_f is the freezing temperature at atmospheric pressure, ΔV is the change in volume on transformation from liquid to solid and H_f is the latent heat of fusion. For magnesium alloys, an applied pressure of 100 MPa produces a 6.5°C increase in freezing temperature.

Because freezing takes place under a high applied pressure, the "castability" of the alloy is of no real concern and both conventional casting alloys and erstwhile wrought alloys can be squeeze cast without difficulty. Because the process can guarantee zero porosity, there would appear to be no logical necessity for NDT inspection and the casting factor should be unity for squeeze cast components. The microstructure of the castings can be controlled by control of the casting variables alone, without recourse to nucleating agents.

MICROSTRUCTURES AND PROPERTIES: CASTING ALLOYS

Casting alloy AZ91

The most common magnesium casting alloy is AZ91, which contains ∿9% aluminium and ∿1% zinc. This alloy is used both as a gravity die casting alloy and as a high pressure die casting alloy in which cases the grain sizes are large and small, respectively, and are not easily controllable. Additionally, both types of castings contain considerable, but non-uniform, amounts of porosity and therefore the recorded mechanical properties are influenced by both grain size and porosity. With squeeze castings, the grain size can be effectively controlled and the true mechanical properties of the full-integrity material can therefore be obtained as a function of grain size.

Figure 2 shows the measured variation of UTS as a function of grain size for the squeeze cast AZ91 alloy in its fully heat treated condition. The grain size effect is very pronounced, the UTS increasing from 170 MPa for a grain size of ∿1.3mm to a value of 260 MPa for a grain size of ∿0.23mm. Over this range of grain sizes the secondary dendrite arm spacing (SDAS) did not vary significantly. Hence these results show a true grain size effect for these castings, from which any influence of SDAS and porosity have been eliminated.

In the absence of porosity the influence of other casting defects,such as oxide inclusions, can be established. Figure 3 illustrates the progressive decrease in UTS brought about by increasing inclusion size. The range of UTS values with zero inclusions is due to the grainsize effect: the value of 260 MPa is the ultimate UTS level for AZ91 for a grain size of ∿ 0.23mm.

Figure 4 shows the variation in properties of AZ91 in the as-cast and in the fully heat treated condition for sand cast, gravity die cast, high pressure die cast and squeeze cast material. In every case the squeeze cast material shows the highest values of UTS, 0.2% proof stress and elongation to failure. The full heat treatment cycle increases the UTS of the squeeze cast material from 200 MPa to 260 MPa with an associated decrease of elongation of only 1%.

Although the squeeze casting process does yield the best mechanical properties of the cast AZ91 alloy, this alloy is inherently a low-strength alloy and the improvements attainable by squeeze casting and fully heat treating are only marginal when compared with gravity die casting. Further improvements can be expected for alloy compositions outside casting specifications which are "difficult to cast". Squeeze casting can then utilize the extra solute content to improve the mechanical properties by enhanced precipitation hardening in the porosity-free matrix.

MICROSTRUCTURES AND PROPERTIES: WROUGHT ALLOYS

Wrought alloys AZ31 and ZCM711

The well-established wrought alloy AZ31 (containing 3% aluminium and 1% zinc) is used in the extruded, work hardened condition, and in the longitudinal direction the alloy attains a UTS value of 255 MPa. The alloy does not respond to heat treatment. The squeeze cast alloy in the as-cast condition attains a UTS value of 195 MPa in specimens having a grain size of ∿0.2mm. This value is less than that of the extruded bar in the rolling direction, but probably better than the properties of the extruded bar in the transverse direction.

Recent alloy developments at MEL have produced a new wrought magnesium alloy ZCM711, which contains 7% zinc, 1% copper and 1% manganese. This alloy does respond to heat treatment. It is used in the wrought and fully heat treated condition and possesses excellent room temperature strength in the longitudinal direction but much poorer properties in the transverse direction. Figure 5 shows the reason for this anisotropy: the alloy contains a high volume fraction of stringers of the intermetallic compound Mg_2Cu. By way of contrast, Figure 6 illustrates the microstructure of the fully heat treated squeeze cast ZCM711 alloy: the structure is fine-grained and equiaxed and the Mg_2Cu intermetallic is evenly distributed throughout.

A comparison of the mechanical properties of the wrought and squeeze cast ZCM711 alloy is given in Figure 7. The UTS and the 0.2% proof stress of the wrought alloy in the longitudinal direction are extremely high compared with the values obtained in the short transverse direction and with the results derived from the squeeze cast material. In contrast, the elongation measured on the squeeze cast alloy, at 16% is more than twice that measured on the wrought material in either direction. The absolute value of the UTS of the squeeze cast material is only 272 MPa, very little different from the figure of 260 MPa obtained from the AZ91 alloy. Such a comparison highlights the fact that the precipitation hardening ZCM711 alloy does not seem to react in the same excellent way as do the aluminium-based precipitation hardening alloys, such as 7010, the squeeze castings of which match the longitudinal properties of the wrought material.

NEW ALLOY DEVELOPMENT

Because the pressures used in squeeze casting affect the melting temperatures of the alloys and the solubilities of solutes in the matrix phase, it is possible, indeed necessary, to design alloys for this new casting process in order to take full advantage of any benefits that might accrue from these changes of phase fields. Hi-Tec Metals R & D Ltd have embarked upon an alloy development programme to tailor alloys for squeeze casting in order to achieve properties suitable for specific applications. One such alloy, designated HTM 1 for the time being, was developed to withstand substantial (50MPa) loading in creep and fatigue at temperatures up to 453K. The only alloys previously available to fulfil these engineering requirements were the MEL alloys QH21, QE22 and WE54 which are very expensive and, for this reason, were unacceptable for this particular application.

Figure 8 shows the UTS values of HTM 1 in comparison with ZCM711 in the longitudinal direction as a function of temperature up to 473K. Although the wrought alloy exhibits better tensile properties at the lower temperatures, the newer alloy HTM 1 maintains its tensile strength to higher levels at the highest temperatures. Also shown on the figure are the room temperature and 523K values of UTS for QH21 and QE22. The creep curves of HTM 1 and ZCM711 are shown on Figure 9 along with creep curves of some other inexpensive magnesium alloys. Although the ZCM711 alloy shows very good tensile strength (230 MPa) at 423K, its creep strength at this temperature under a load of 100 MPa is very poor. On the other hand, the tensile strength of HTM 1 at 423K is only 200 MPa but its creep strength under the same load of 100 MPa is extremely good.

The fatigue properties of HTM 1 at room temperature and at 453K are given in the Table in terms of the load capability at specific cycles to failure.

	5×10^4	5×10^5	5×10^6	10^7
300K	148.1	115	100	98.7
453K	129.7	90.2	73.3	69.8

These figures indicate the excellent fatigue strength of this alloy, which, coupled with the good creep behaviour, illustrate the unique properties of this new magnesium alloy for the medium temperature range of applications.

MAGNESIUM-BASE METAL MATRIX COMPOSITES

There is no practical advantage in attempting to fibre-reinforce castings if they contain any porosity, since the porous regions will control the mechanical response to the applied stresses. It is only when porosity-free castings are available via the squeeze casting route that fibre-reinforcement becomes a practical proposition.

In the present series of experiments at Hi-Tec Metals R & D Ltd several different types of fibres have been encapsulated in different metallic matrices. In the results reported here the matrix is, in every case, AZ91. The fibres used have been Saffil[R], a glass/graphite mixture, and stainless steel wire.

The Saffil[R] used was in the form of hard, pre-fired mats having densities of 0.44 - 0.95 gms/cc the alumina fibres being 14μm in diameter. The glass/graphite pads were rather more open than the Saffil[R] pads and produced composites having ∿9vol% graphite: the diameter of the graphite fibres was 8μm. The stainless steel wire was 50μm in diameter and was aligned to give a packing density of ∿60vol%. Despite the close packing of the fibres no problems were encountered in producing the metal matrix composites, since the applied pressure was always sufficient to squeeze the liquid magnesium alloy through the fibre bundles and around the individual fibres to preclude any gas entrapment in the fibre mat and to ensure absolute metal-to-metal interfacial contact. This was true for all the different types of fibre arrays. Figure 10 shows the Saffil[R] fibres in a fibre pad before infiltration and Figure 11 illustrates the perfect infiltration observed in the metal matrix composite. Figures 12 and 13 illustrate the same features for the glass/graphite pad, again showing total infiltration of the fibres. Although the temperature of the magnesium was sufficiently high to melt the fine glass fibres during infiltration, the graphite fibres remained fully dispersed, as is shown in Figure 13. To complete the picture of the microstructural character of the metal matrix composites, Figure 14 illustrates the high packing density of the unidirectionally aligned stainless steel wire and again shows complete penetration of the inter-fibre spaces by the liquid magnesium alloy.

The simplest mechanical property to measure on composite materials is their hardness. Figure 15 shows the hardness of two Saffil[R] containing composites and one glass/graphite composite in comparison with the hardness of the matrix phase alone. It is noteworthy that the glass/graphite composite, containing only ∿9% of graphite, has a hardness very similar to that of the 25 vol % Saffil[R]-reinforced alloy: this is a reflection of the higher modulus of graphite compared with alumina. It is also interesting that the temperature dependence of the hardness values is the same for all three composites, implying that the load transfer characteristics are common to all three and that the progressive decrease observed is due entirely to the softening of the magnesium matrix alloy.

Creep and fatigue curves have been obtained on the 16vol% Saffil[R]-reinforced alloy and compared with AZ91 tested under the same conditions. The results are shown in Figures 16 and 17 and indicate that the composite material has a creep life of an order of magnitude better than the matrix phase and a fatigue strength double that of AZ91 alloy alone. This latter result is similar to that already found for aluminium alloy composites containing Saffil[R] pads. Because the fracture toughness values for the high-hardness composites are very low at room temperature (in the region of 10 MPam$^{\frac{1}{2}}$) these metal matrix composites will only find a market in high temperature applications where their superior creep and fatigue properties can be fully utilized without fear of an attendant low fracture toughness.

SUMMARY AND CONCLUSIONS

The squeeze casting process is now being used commercially for the production of components for the automotive industry manufactured in both aluminium- and magnesium-base alloys and as monolithic castings and metal matrix composites. The economics of the process have been shown to be favourable in comparison with forging and the superior quality of squeeze castings has already displaced gravity die casting for some applications. Rather complex-shape castings have already been produced via the squeeze casting route but the intricacy of high pressure die castings has not yet been attained, although it should not be unattainable.

The primary advantage of squeeze castings is their high quality. The casting process guarantees perfect replication of all details of the die form and the casting itself is porosity free. Thus, the castings should rate a casting factor of unity and the need for NDT inspection should be obviated. The microstructure of the squeeze cast material is fully controllable by control of the casting parameters and magnesium alloys having a uniform grain size of ∿200μm are readily achievable. But because the casting conditions also affect the equilibrium diagram in addition to producing zero porosity castings, it is possible to squeeze cast alloys outside the normal specification ranges and thus formulate new alloys for this new casting process which have specifically required properties. The new alloy HTM1 referred to in Figures 8 and 9 was chemically formulated and microstructurally designed for good creep and fatigue behaviour up to 453K and squeeze castings of this alloy are now being tested for commercial application.

The second major advantage of the squeeze casting process is the ease with which metal matrix composites can be prepared. Fibre preforms can

be placed in the die and incorporated into the
squeeze castings in the exact positions where
strengthening is required. The pressures applied
during the casting procedure are sufficient to
guarantee perfect infiltration of the liquid metal
into every interfibre recess and the interfacial
contiguity is better than with other low pressure
infiltration techniques. Magnesium-graphite
composites made via the squeeze casting route are
shown to have excellent mechanical properties.

Squeeze casting is considered to be an ideal way
to produce net-shape and near-net-shape components
for the automotive and aerospace industries.
Several magnesium castings have been prototyped
and tested in readiness for industrial exploita-
tion. Similarly, it is easy to fabricate metal
matrix composites by squeeze casting and it is
considered that this technique provides the most
effective and efficient route to produce composite
components for engineering applications.

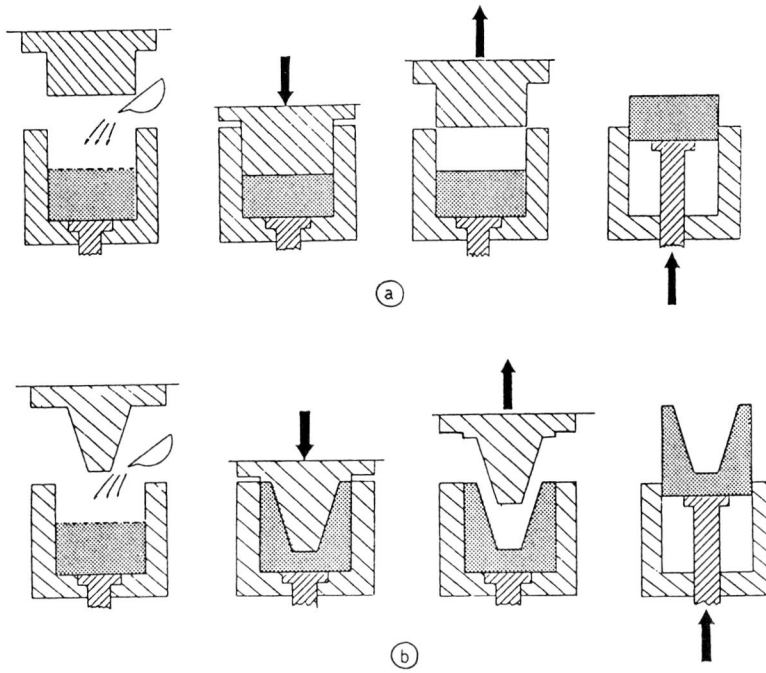

Fig.1. Schematic of squeeze casting process: (a) solid ingot production; (b) hollow shapes.

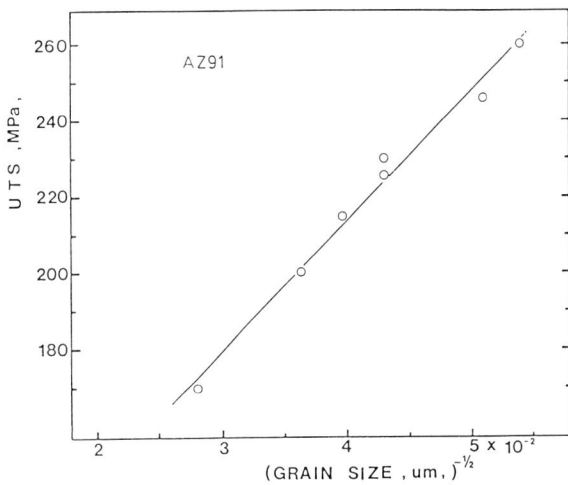

Fig.2. UTS of squeeze cast AZ91 as a function of grain size.

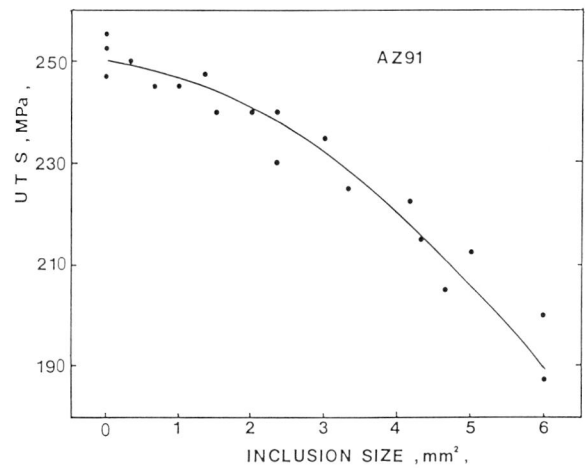

Fig.3. UTS of squeeze cast AZ91 as a function of inclusion size.

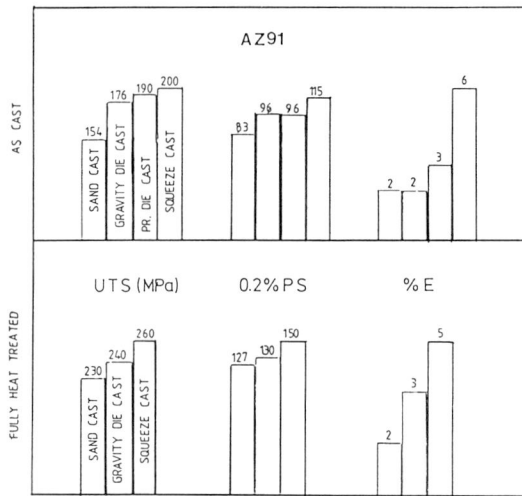

Fig.4. Mechanical properties of squeeze cast and fully heat treated AZ91 as a function of the casting process.

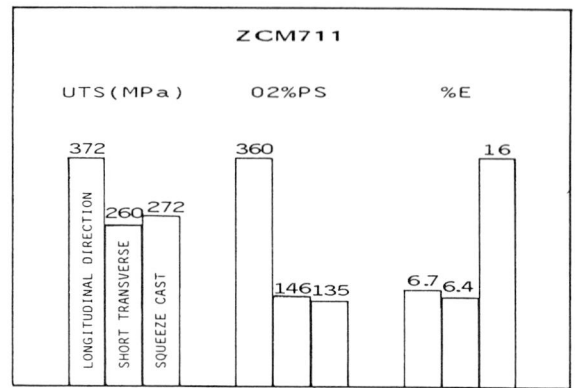

Fig.7. Mechanical properties of ZCM71 in wrought form compared with squeeze cast material, all in fully heat treated condition.

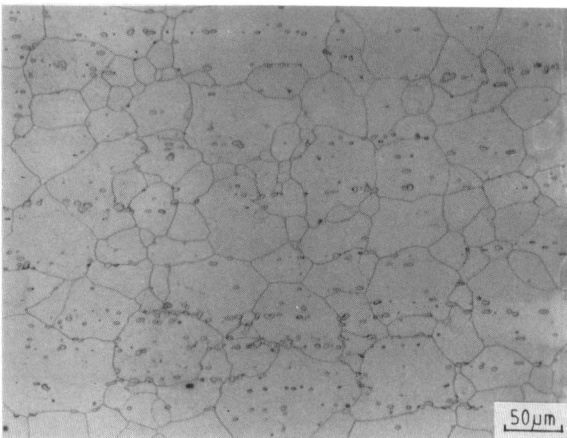

Fig.5. Microstructure of ZCM711 in the longitudinal section showing stringers of Mg$_2$Cu compound.

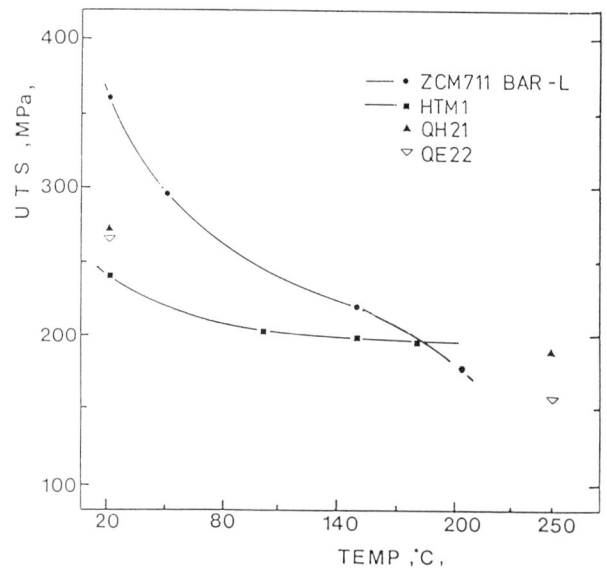

Fig.8. Comparisons of the UTS values of HTM 1 and ZCM711 as a function of temperature.

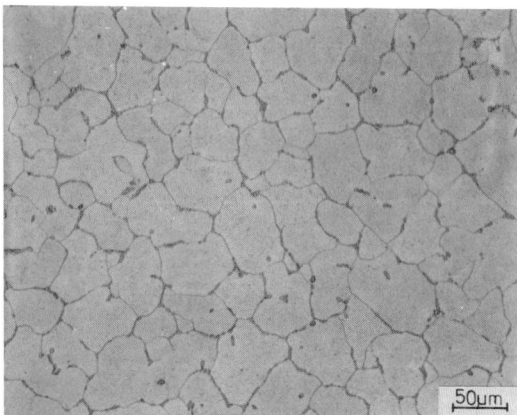

Fig.6. Microstructure of squeeze cast ZCM711 showing uniform, grain boundary distribution of Mg$_2$Cu compound.

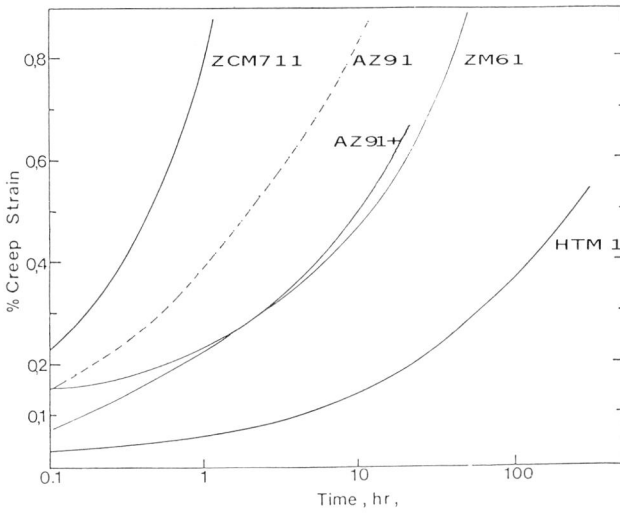

Fig.9. Creep curves of several wrought and cast
magnesium alloys showing the excellent
creep behaviour of HTM 1. Test
temperature 423K, load 100MPa.

Fig.10. SEM of Saffil[R] fibres in rigid pad.

Fig.11. Composite of Saffil[R] and AZ91 showing
complete infiltration of pad by liquid metal.

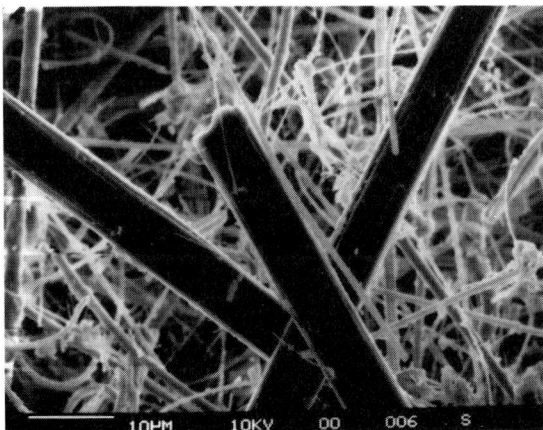

Fig.12. SEM of glass/graphite pad, the graphite
fibres being ⌄8µm in diameter.

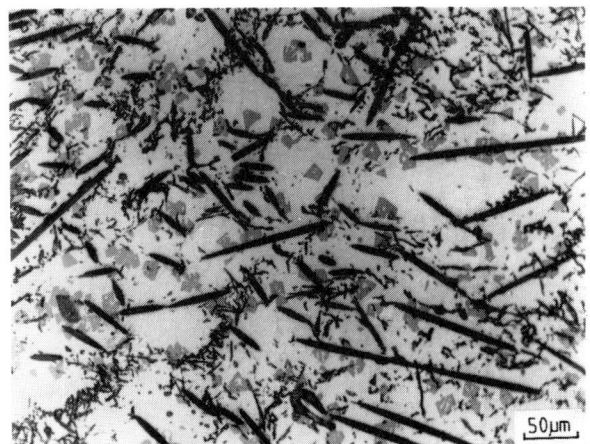

Fig.13. Composite of glass/graphite and AZ91.
The glass fibres have melted during
liquid metal infiltration and generated
large crystals of Mg_2Si.

81

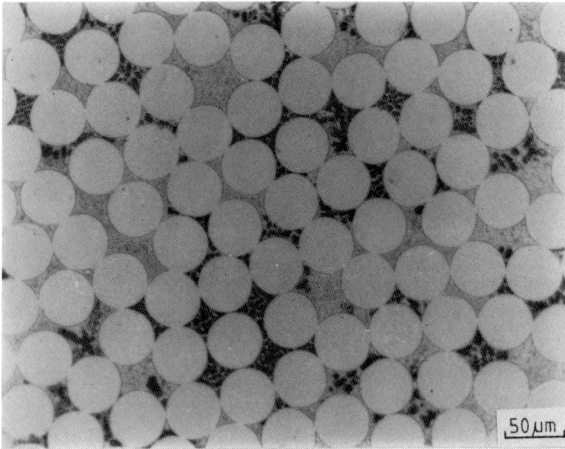

Fig.14. Composite of stainless steel fibres in an AZ91 matrix.

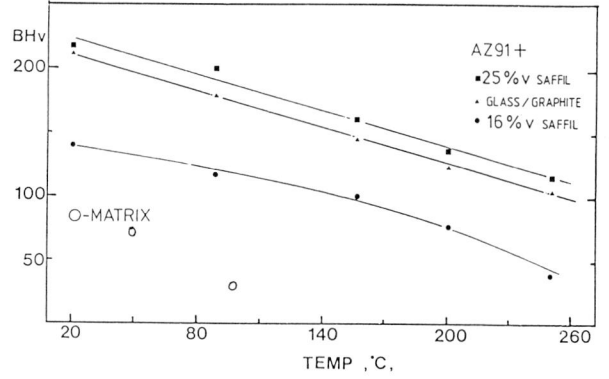

Fig.15. Hot hardness values of AZ91-based MMCs containing Saffil[R] and glass/graphite, in comparison with matrix phase alone.

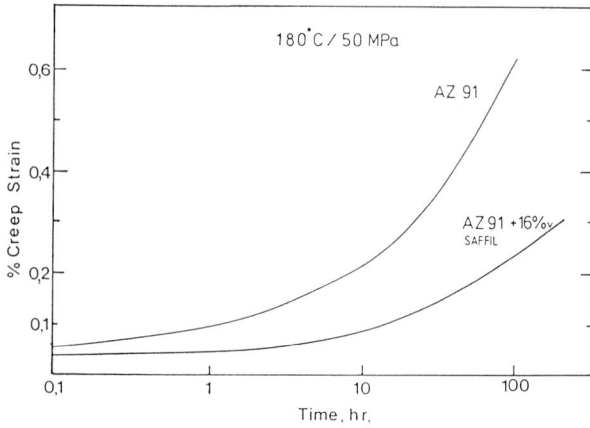

Fig.16. Creep curves of AZ91 and MMC of AZ91 containing 16 vol.% Saffil[R].

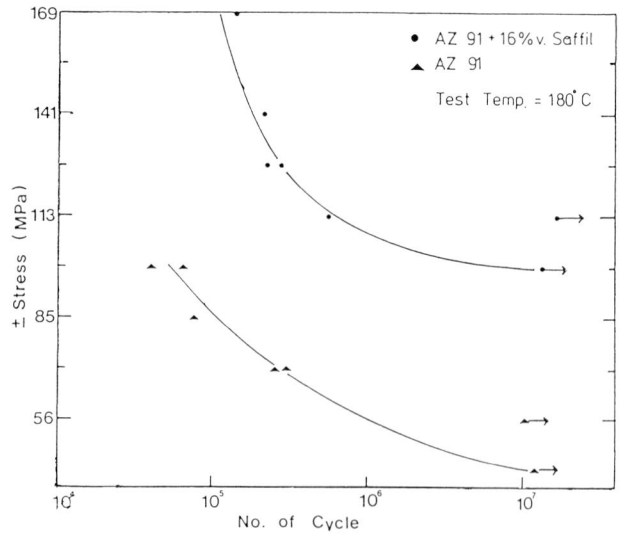

Fig.17. High temperature fatigue curves of AZ91 and MMC of AZ91 containing 16 vol.% Saffil[R].

Rapid solidification processing
of magnesium alloys

F HEHMANN and H JONES

*Dr Jones is in the Department of Metallurgy
at the University of Sheffield, England.
Mr Hehmann is with the Max Planck Institut
für Metallforschung, Stuttgart, West Germany.*

SYNOPSIS

Although magnesium alloys were among the first
(c.1950) to be subjected to rapid solidification
processing (RSP) with a view to enhancement of
their engineering properties, the major research
effort expended on RSP since the 1970's has left
them practically untouched. The present contribution considers the potential of RSP as a route for
development of improved metallic materials based
on magnesium, reviews recent published work
directed towards that objective and discusses
recent results obtained by the authors at
Sheffield concerned with the effect of RSP on
magnesium alloys containing Y and/or lanthanide
additions.

1. INTRODUCTION

Rapid solidification processing (RSP) embraces the
family of production processes now available that
secure movement of the solidification front at
velocities that can reach tens of metres per
second. Such extreme conditions of solidification
can result in large departures in alloy constitution (extended solid solutions and nonequilibrium
crystalline, quasicrystalline or glassy phases) as
well as substantial refinement in scale of the
solidification microstructure. The enlarged scope
for alloying and novel microstructures made acces-
sible by RSP has established a new branch of
materials science with an attendant panoply of
new materials with novel or enhanced combinations
of chemical, electrical, magnetic and mechanical
properties [1,2].

Rapid movement of a solidification front
requires either (i) a short path for external
extraction of the latent heat of solidification it
releases or (ii) a large undercooling of the melt
prior to solidification so that solidification
front velocity is not limited by the rate at which
latent heat can be extracted externally during
solidification. Because presence of inoculants
generally frustrates attempts to achieve large
prior undercoolings in extensive undivided volumes
of melt, condition (ii) also, in practice dictates
a short solidification path. The highest attainable front velocities require this path to be
shorter than 1μm, though the range 10 to 100μm is

more typical of most   practical methods of rapid
solidification now in use.

The present contribution will first summarize
reported experience to date on the applicability
of available methods for RSP of magnesium-based
materials and their effects, will go on to identify some opportunities for further investigation
and conclude by reviewing some recent work including some carried out by the authors at Sheffield
on magnesium alloys containing yttrium and/or
lanthanide additions.

2. APPLICABILITY OF METHODS OF RSP TO MAGNESIUM ALLOYS AND THEIR EFFECTS

Methods of RSP of melts divide broadly into three
categories (Figure 1): (1) those involving spray
or droplet formation, such as atomization into
powder or flake particulate (2) those derived
from continuous chill casting, such as melt-
spinning of ribbon or thin strip by melt-stream
impingement on to a rotating chill block, and (3)
those featuring in-situ rapid melting and resoli-
dification in intimate contact with a chill surface.

Examples of all three categories of method
have been applied with success to magnesium
alloys (Table 1). In category (1), gas-blast [3-
5] and then rotary [6-9] atomization was applied
by Dow Chemical Co. in the late forties and early
fifties to generate magnesium alloy powder particulate in up to 1000 lb batches for consolidation
by extrusion into test-pieces and ultimately
engineering parts with improved properties.
Later work [10] employed the rotating electrode
process to generate magnesium alloy shot particulate for the same purpose. The more powerful
gun technique of splat quenching was first
applied to magnesium alloys in the sixties and
was effective in extending solid solubility by a
factor of 2 from the maximum equilibrium level of
11.8at% to 22at%Al in Mg [11]. The related cata-
pult technique produced solid solubility extension by 2.5 times to 2.5at%Mn in Mg [12] and
later work generated metastable fcc phases in
the Mg-Sn [13] and Mg-Pb systems [14] and metallic glasses in the Mg-Ni [15,16] and Mg-Cu [15-
18] systems. Levitation melting and two-piston
splat-quenching has been used by us at Sheffield
to prepare rapidly-solidified samples of a wide
range of magnesium alloys.

In category (2), the use of free-jet chill-block melt-spinning to produce continuous ribbon of a magnesium alloy was first reported by Calka et al in 1977 [19]. The resulting Mg-Zn metallic glass was the first to be formed between two 'simple' metals (as opposed to between two transition metals or between a transition metal and a metalloid). Numerous publications followed and continue to appear (see [20] for references) characterizing this novel material and its derivatives, primarily concerned with their atomic and electronic structure and physical properties and those of the related Mg-Cu and Cu-Mg glasses. The pulverization of melt-spun ribbon [21-23] or chopping of twin-roll quenched flake [24] has been employed latterly to generate RS particulate in bulk for consolidation by extrusion into novel wrought engineering alloy compositions. Even one of the newest processes, melt-overflow, has been applied with good effect to magnesium alloys [25].

Turning to category (3), this has received relatively less attention to date for magnesium alloys, reflecting the fact that this group of techniques has received major attention only in the present decade. Kattamis [26], however, reported in 1981 on the effect of laser and electron beam surface melting on the microstructure and corrosion behaviour of the high strength wrought ZK60 magnesium alloy (Mg-6wt%Zn-0.7wt%Zr) and Kalimullin and Kozhevnikov [27] reported in 1985 that laser surface treatment increases substantially the corrosion resistance of the Mg-Li base alloy MA21 (Mg-8wt%Li-5wt%Al-4wt%Cd-1wt%Zn-0.4wt%Mn) in 3% NaCl.

Thus, methods from each of the three main categories of rapid solidification from the melt have been usefully applied to magnesium alloys, and consolidation of derived particulates has been successfully carried out by compaction and extrusion down to ambient [3,6,9,10,21-24]. Small-sample splat-making or chill-block melt-spinning seem equally suitable for initial monitoring of the effect of rapid solidification on experimental alloy compositions with atomization and planar-flow casting both established as viable methods for generating larger quantities of particular alloys for evaluation in bulk or consolidated form.

3. IDENTIFICATION OF POSSIBILITIES FOR FURTHER INVESTIGATION.

Magnesium alloys have to date found only limited application as engineering materials in spite of large reserves (third most abundant metal after aluminium and iron), costs comparable with aluminium and the lowest density of all the engineering metals. Limitations include formability and strength associated with the restricted plasticity of the cph crystal structure of magnesium at normal temperatures and corrosion resistance associated with the inability of known magnesium alloys to form a self-healing protective film comparable in effect to alumina on aluminium alloys or chromia on stainless steels. Rapid solidification provides a number of possibilities to help overcome such limitations. These will be considered in turn.

3.1 Extension of solid solubility

The maximum equilibrium solid solubility C_{max}^{eq} is a major factor limiting the choice and level of wrought alloy additions. Available data [28-30]

identifies some 24 alloy elements as exhibiting potentially useful levels of equilibrium solid solubility ($C_{max}^{eq} > 1$at%) in magnesium (Table 2). All but 2 (Sc and Ga) of the 9 with lowest atomic number feature prominently in established engineering magnesium alloys [20]. All but 2 (Mn and Yb) of the 24 solid-soluble elements have atomic sizes [31] within the $\pm$ 15% limit for extensive solid solubility (most are within $\pm$ 12% of the atomic diameter of magnesium) and those with the smallest size differences tend to give higher solubilities. Figure 2 shows atomic radius r as a function of atomic number z with $\pm$ 7½% and $\pm$ 15% 'limits' superimposed. Experience with other systems, such as those based on aluminium [32], suggests that extensive solid solubility is unlikely to result from rapid solidification of the melt for alloy additions outside the 15% limit, but that substantial extensions are possible for additions with atomic sizes within the limit. This more than doubles the number of elements that could exhibit useful solid solubility compared with equilibrium conditions. To date only three elements have been reported to exhibit solid solubility extension in magnesium as a result of rapid solidification from the melt i.e. Al, Mn and Ga, leaving ample scope for further exploration.

One benefit of such investigation would be to identify elements that extend solid solubility sufficiently to reduce the $^c/_a$ ratio of the cph lattice of Mg to below the level at which non-basal slip becomes active at normal temperatures with associated enhancement of deformability. Additions known to reduce $^c/_a$ of Mg are limited at present to Li and Ag of which Li initiates non-basal slip at Li contents above 8 at%, at which the $^c/_a$ ratio has decreased to 1.618, with a significant increase in ductility [33]. Extension of solid solubility of Ag in Mg from its C_{max}^{eq} of 3.8at% to some 7.6at% by rapid solidification would be required to reduce $^c/_a$ to this level in Mg-Ag alloys. Since $^c/_a$ is primarily a function of valency electron to atom ratio $^e/_a$ in magnesium alloys, with $^c/_a$ reducing with decreasing $^e/_a$ [34], the largest rate of reduction is produced by elements of lowest (preferably zero) valency. Unfortunately most of the more solid soluble additions to Mg have valencies equal to (Zn,Cd) or higher (Al,Ga,In,Sn,Pb) than Mg so maintain or increase $^c/_a$ rather than decreasing it.

3.2 Formation of new alloy phases

In principle, such new phases could be crystalline, quasicrystalline or glassy. New crystalline phases reported to have been formed by rapid solidification from the melt include the fcc phases referred to in Table 1 for the Mg-Sn and Mg-Pb systems but both decompose rapidly at room temperature and contain too much Sn and Pb to retain the density advantage of magnesium over, say, aluminium alloys. It is not clear what advantage the anti-PbCl$_2$-structured form of Mg$_2$Si reported [35] to form in Mg-Si alloys would have over the equilibrium form which serves as a very effective thermally stable dispersoid in rapidly-solidified magnesium alloys. The only Mg-containing quasicrystalline (five-fold symmetry) phase so far reported has it only as a minor component [36] and density-increasing additions such as Zn,Cu or Ni need to be present at levels of at least 10 to 20 at% to generate Mg-based glasses under standard conditions of rapid solidification (splats or melt-spun ribbons 30 to 50μm in thickness) [20].

Attempts have been made to form such glasses at lower contents of Cu with the aid of a low-density ternary addition such as Ca [37]. None of these glasses is particularly stable thermally, decomposing in the range 350 to 490 K on heating [20] and, for the Mg-30at%Zn glass, within six years at room temperature [38]. A suitably-controlled thermal decomposition of suitable metallic glasses can, however, result in microcrystalline products with very attractive properties [39], in the same way that controlled thermal ageing of a supersaturated crystalline solid solution can result in useful properties (tempering of ferrous martensites; age-hardening of nonferrous alloys). This possibility of developing useful engineering properties by controlled devitrification of Mg-based metallic glasses is one of several avenues that have yet to be explored.

3.3 Microstructural refinement and dispersoid formation.

Constitutional effects of rapid solidification as embodied in extension of solid solubility and formation of new phases are system-specific so are highly sensitive to alloy composition range as well as to the thermal conditions of rapid solidification imposed. Microstructural refinement, in contrast, is a universal consequence of rapid solidification, and refines grain and solidification cell size of the matrix phase as well as the size and spacing of any dispersion of second phase particles resulting from the solidification process. Homogenization treatments that may be only partially effective for ingot-derived materials can thus be curtailed or even eliminated entirely for the rapidly-solidified state. Published reports of the effect of rapid solidification on microstructure of magnesium alloys are surprisingly limited. Dendrite cell sizes ∿ 10μm were obtained by Busk and Leontis [3] in ∿ 200μm diameter gas-atomized AZ31 (Mg-3wt%Al-1wt%Zn-0.2wt%Mn) magnesium alloy powder while cell sizes ∿ 5μm were shown by Busk [6] for 100μm-diameter rotating-disc atomized ZK60 magnesium alloy particulate, compared with ∿ 60μm for ingot-material cast by the direct-chill process (Figure 3). This order of magnitude refinement in the atomized ZK60 is reflected in much improved uniformity of microstructure in extrusions made from the atomized particulate compared with extrusions of ingot material (Figure 4). These powder extrusions exhibited 15 to 40% higher compressive yield strength than ingot extrusions of the same ZK60B alloy (Mg-6wt%Zn-0.6wt%Zr) and led to military applications as support beams for the floor and built-in loading ramp of the C-133 transport aircraft (Figure 5) [9].

The further microstructural refinement that becomes possible by using substrate quenching techniques of RSP has been demonstrated only recently for magnesium alloys. Meschter and O'Neal [24] reported α-plate refinement by a factor as large as 30 in twin-roll rapidly-solidified flake of Mg-Li base alloy compared with chill-cast material. Cell size as small as 0.4μm was reported by Skjerpe [40] for melt-spun ribbon of Mg-2wt%Mn alloy (an order of magnitude finer than would be typical of coarse atomized powder particulate). Grain size was as small as one-twentieth that of cast material. Cell sizes of 0.1 to 0.3μm / grain sizes of 0.4 to 1.5μm have been reported by Chang et al [21-23] for

melt-spun Mg-Al and Mg-Zn-Al base alloys containing further additions of Si, Y or lanthanides. Grain sizes as small as 0.7μm were retained after extrusion in alloys containing a suitably stable second phase present as a finely distributed dispersoid as a result of RSP. Chang et al showed that such refinements of microstructure can lead to notable improvements in strength and corrosion resistance in extruded material compared with the best wrought ingot alloys of the ZK60 and AZ91 HP type (Mg-9.5wt%Al-0.5wt%Zn-0.3wt%Mn). Extrapolation of the power relationship* between dendrite cell size d and local solidification time t_F established by Kattamis et al [41] for Mg-5wt% Zn indicates local solidification times in the range 10^{-4} to 10^{-5} s for the cell sizes of 0.1 to 0.3μm obtained by Chang et al by chill substrate rapid solidification. These values of t_F compare with 0.1 to 1 s for d of 5 to 10μm typical of coarse atomized powder and several minutes for d of 100μm in conventionally cast material. The fine and relatively stable grain size afforded by RSP should be valuable both in improving ductility and strength of magnesium alloys at normal temperatures and in enhancing superplastic behaviour, well-established in magnesium alloys, at elevated temperatures. A key requirement in selecting a second phase dispersoid suitable for retaining the fine grain size obtained by RSP, is a low solid solubility in magnesium. Figure 6 shows the relationship between melting point and solid solubility of dispersoids in magnesium first noted by Carapella [42]. Two of the least soluble dispersoids Mg_2Si [21-23,43,44] and Mg_2Ba [43], as well as one not quite so insoluble, Mg_xCe [22, 23,45] and related lanthanide phases [22,23], have already been shown to be effective both in retaining grain size [21,23] and in improving thermal stability [21-23,45] and hot strength [43,45] in several Mg-based compositions made via RSP. Such effects of Y and lanthanide additions in magnesium alloys when subjected to RSP are the subject of the next section.

4. RAPID SOLIDIFICATION OF MAGNESIUM ALLOYS CONTAINING YTTRIUM OR LANTHANIDE ADDITIONS.

The effect of yttrium amd lanthanide additions in improving the elevated temperature properties of magnesium alloys has led to the development of a number of commercial alloys including the creep resistant casting alloys EZ33 (Mg-3wt%RE**-2.5wt% Zn-0.6wt%Zr) and most recently WE54 (Mg-5.25wt%Y-3.5wt%RE**-0.5wt%Zr). The tensile properties of WE54 in the range 150 to 300°C are reported to be superior to those of any available aluminium-based casting alloy. Yttrium and the lanthanides all exhibit eutectic terminal equilibria with magnesium, the phases involved being αMg and an intermetallic compound usually with the stoichiometric $Mg_{12}X$, (X = Ce,Pr,Nd) or $Mg_{24}X_5$ (X = Y,Tb, Dy,Ho,Er,Tm or Lu). Maximum equilibrium solid solubilities are reported to be small (< 0.1at%) for La,Ce and Pr increasing to several atomic per cent for Y and the heavy lanthanides (Tb,Dy,Ho,Er, Tm or Lu). This difference at least partially reflects a reduction in atomic radius of the lanthanide giving borderline Hume-Rothery size fac-

---

* Footnote: $d = 10.5t_F^{0.4}$ where d is in μm and t_F is in seconds [41].

**Footnote: RE = rare earth i.e. lanthanide

tors in Mg of 17.3, 14.0 and 14.1% for La,Ce and Pr compared with increasingly favourable values (12.5,11.4,10.7,10.3,9.7,8.9 and 8.3% respectively) for Y.Tb,Dy,Ho,Er,Tm and Lu (atomic size data from [16]).

The effect of lanthanide additions on rapidly solidified magnesium alloys was included in the early Dow work on extrusions of gas atomized particulate [3,44,46,47] and resulted in development of alloy ZE62 (Mg-6wt%Zn-2wt%MM+-0.5wt%Zr) with increased strength, stability and weldability compared with ZK60B [6]. The work of Meschter and O'Neal [24] on dispersoid formation in twin-roll rapidly-solidified Mg-9wt%Li alloy included the addition of Ce. Extrusions of the alloy with a 1.7wt%Ce addition showed a 50% increase in yield strength at 150°C compared with the identically-processed Ce-free Mg-8.5wt%Li composition [45]. Chang et al [22,23] investigated the effect of 0.5 to 2at%Y,Ce,Pr or Nd additions on melt-spun Mg-Zn-Al,with and without an Si addition,that was pulverized prior to consolidation by extrusion. The resulting $\sim$ 2 vol% of Mg_3Ce, Mg_3Pr, Mg_3Nd or $Mg_{17}Y_3$ dispersoid in the melt-spun Mg-Zn-Al(-Si) showed negligible coarsening during extrusion compared with MgZn or $Mg_{17}Al_{12}$ precipitates. The high levels of strength (Y.S. up to 476MPa and UTS up to 576MPa) for these alloys at room temperature, fall to two-thirds of their room temperature values on testing at 100°C and drop to one-third or one-quarter of room temperature values on testing at 150°C. These reductions in strength are accompanied by 10- and 40-fold increases in elongation to fracture at 100 and 150°C respectively, strength levels at 150°C being comparable with wrought ingot alloys ZK60 and AZ91HP. Corrosion in 3%NaCl solution at 25°C for the Mg-Al-Zn-Y alloy,at 11 mdd, however, was only one-sixth that of AZ91HP-T6 and one-twelfth that of ZK60A-T5 (Mg-5.5wt%Zn-0.5wt%Zr). A previous publication by us [20] reported on the effect of RSP on four engineering magnesium alloys including two containing lanthanides (ZE41 = Mg-4.2wt%Zn-1.3wt%RE-0.7wt%Zr and EZ32 = Mg-2.7wt%RE-2.2wt%Zn-0.6wt%Zr). We conclude this present paper with some corresponding results obtained for the binary alloys Mg-3 and 13wt%Ce, Mg-3.5,11 and 17wt%Nd and Mg-7.5,10,14 and 23wt%Y, the ternary alloy Mg-18wt%Y-6wt%HRE and the commercial alloy WE54.

RSP was carried out, as previously, by the two-piston technique in an argon atmosphere to give splats $\sim$ 100μm in thickness and between 15 and 30mm in diameter. Microhardness measurements were carried out at 15g load on polished cross-sections in both the as-splatted condition and after heat treatment for 1h at temperatures between 100 and 400°C. Figure 7a-c gives microhardness as a function of alloying content for the three binary systems and shows that Ce is the most effective hardener and that RSP by the twin piston technique gives an increment in hardness typically $\sim$ 30% compared with the chill-cast condition for both Ce concentrations and for the two higher concentrations of Nd. Results for microhardness after 1h exposure at temperatures up to 400°C are plotted in Figures 8a-d for the three binary and two more complex alloys for the twin-piston-splatted and chill-cast conditions. Figures 8a-d show that the increment in hardness

given by RSP except for Mg-Y is maintained after treatment at all temperatures up to 400°C. Age-hardening responses are recorded for the Mg-13wt% Ce in the RSP condition at $\sim$ 100°C and for both conditions at $\sim$ 300°C (Figure 8a). This higher temperature treatment resulted in a hardness level of 200 kg/mm^2 after 1h at 400°C for rapidly solidified Mg-13wt%Ce, a remarkable result for an Mg-based alloy, although the splats of this alloy were very brittle. Rapidly-solidified Mg-11wt%Nd showed a somewhat less marked effect at $\sim$ 200°C (Figure 8b), while rapidly-solidified Mg-23wt%Y showed a peak hardness of 180 kg/mm^2 after 1h at 250°C (Figure 8c). This peak reached 200 kg/mm^2 after the same treatment for the rapidly-solidified Mg-18wt%Y-6wt%HRE++ alloy and maintained this value even after 1h-treatment at 400°C (Figure 8d). The WE54 alloy showed some age-hardening response at 150 to 250°C for both the rapidly-solidified and chill-cast conditions. The splats of this alloy were very ductile in bending.

Metallographic studies indicated generally cellular microstructures for the alloys in the as-splatted condition with more evidence of columnar directionality in relation to the chill surfaces for the more dilute Mg-Ce and Mg-Nd alloys and for the Mg-Y alloys. Figures 9a and b show that such microstructures remain essentially intact after 1h at 300°C for Mg-3wt%Ce and after 1h at 200°C for Mg-7.5wt%Y.

The observation of age-hardening responses in rapidly-solidified Mg-Ce and Mg-Nd alloys at levels of alloying well beyond the maximum equilibrium solid solubility is clear evidence that extension of solid solubility was achieved by the conditions of RSP employed. Subsequent X-ray diffraction work at Stuttgart has confirmed that solid solubility extension can reach 1at%(5.5wt%) Ce and 1.4at%(7.8wt%)Nd in Mg by two-piston splat quenching and that even higher levels of extension are attainable under even more extreme conditions of rapid solidification [48]. Peak age-hardening has been reported to occur in 4 to 20h at 200°C in solution-treated and quenched Mg-1.3wt%Ce [49] and Mg-5 to 12wt%Sm [50] alloys, possibly corresponding to the peak at 100°C for rapidly-solidified Mg-13wt%Ce in Figure 8a and that at 200°C for rapidly-solidified Mg-11wt%Nd in Figure 8b. The secondary peak at $\sim$ 300°C for rapidly-solidified Mg-11wt%Nd (Figure 8b) and the peaks at 250°C for rapidly-solidified Mg-23wt%Y (Figure 8c) and Mg-18wt%Y-6wt%HRE (Figure 8d) may have a different origin.

5. CONCLUSIONS

1. State-of-the-art rapid solidification techniques can be applied successfully to magnesium alloys and systematic programmes taking advantage of more effective substrate quenching or equivalent methods are now in progress at several centres in Europe and in N. America.

2. Magnesium alloys present ample opportunities for solid solubility extension, formation of new phases and refinement of microstructure which have hardly begun to be explored.

+Footnote: MM = mischmetal = Ce-26La-19Nd-6Pr

++Footnote: HRE = Heavy rare earth

3. Rapid solidification of magnesium alloys containing lanthanides and yttrium can result in hardening levels and thermally-stable age-hardening effects not achieved in chill-cast or ingot material, attributable to solid solubility extension and/or microstructural refinement in the rapidly-solidified material.

ACKNOWLEDGEMENTS

The new results presented in §4 were obtained as part of research contract SUMAC 37401 carried out by the authors for Magnesium Elektron Ltd at the University of Sheffield between October 1984 and May 1985. The authors are grateful to Magnesium Elektron for permission to publish these results and for their interest and comments.

REFERENCES

1. H. JONES, Rapid Solidification of Metals and Alloys, The Institution of Metallurgists, Monograph No. 8, London, 1982.

2. H. JONES, J. Mater. Sci., 1984, 19, 1043-1076.

3. R. S. BUSK and T. E. LEONTIS, Trans. AIME, 1950, 188, 297-306.

4. D. S. CHISHOLM and R. S. BUSK, Brit. Patent 662312 (1951); US Patent 2630623 (1953).

5. D. S. CHISHOLM, US Patent 2676359 (1957).

6. R. S. BUSK, Light Metals, 1960, 23 (266), 197-200.

7. N. R. COLBRY and G. F. HERSHEY, US Patent 2699576 (1955); Brit. Patent 746301 (1956).

8. D. S. CHISHOLM and G. F. HERSHEY, US Patent 2752196 (1956); Brit. Patent 783685 (1957).

9. G. S. FOERSTER and H. A. JOHNSON, Product Engng., 1958, 29 (19), 80-81.

10. S. ISSEROW and F. J. RIZZITANO, Internat. J. Powd. Met. and Powd. Technol., 1974, 10, 217-227.

11. H. L. LUO, C. C. CHAO and P. DUWEZ, Trans. Met. Soc. AIME, 1964, 230, 1488-1490.

12. N. I. VARICH and B. N. LITVIN, Phys. Met. Metallogr., 1963, 16 (4), 29-32.

13. H. ABE, K. ITO and T. SUZUKI, Jap. Inst. Met., 1970, 11, 368-70.

14. H. ABE, K. ITO and T. SUZUKI, Acta Met., 1970, 18, 991-994.

15. F. SOMMER, G. BUCHER and B. PREDEL, J. Physique, 1980, 41, CB-563 to -566.

16. F. SOMMER, M. FRIPAN and B. PREDEL, in 'Rapidly Quenched Metals', eds. T. Masumoto and K. Suzuki, The Japan Inst. Metals, Sendai, Japan, 1982, pp. 209-212.

17. F. SOMMER, Z. Metallkunde, 1981, 72, 219-224.

18. F. SOMMER, H. HAAS and B. PREDEL, in 'Phase Transformations in Crystalline and Amorphous Alloys', ed. B. L. Mordike, Deutsche Gesellschaft fur Metallkunde, Oberursel, 1983, pp. 95-112.

19. A. CALKA et al, Scripta Met., 1977, 11, 65-70.

20. F. HEHMANN and H. JONES, in 'Rapidly Solidified Alloys and Their Mechanical and Magnetic Properties', eds. B. C. Giessen, D. E. Polk and A. I. Taub, Material Research Society, Pittsburgh, Pa., 1986, pp. 259-275.

21. S. K. DAS and C. F. CHANG, in 'Rapidly Solidified Crystalline Alloys', eds. S. K. Das et al, The Met. Soc. of AIME, Warrendale, Pa., 1985, pp. 137-156.

22. C. F. CHANG, S. K. DAS and D. RAYBOULD, in 'Rapidly Solidified Materials', eds. P. W. Lee and R. S. Carbonara, Amer. Soc. for Metals, Metals Park, Ohio, 1986, pp. 129-135.

23. C. F. CHANG, S. K. DAS, D. RAYBOULD and A. BROWN, Metal Powd. Report, 1986, 41 (4), 302-308.

24. P. J. MESCHTER and J. E. O'NEAL, Met. Trans. A, 1984, 15A, 237-40 and in press.

25. T. O. GASPAR, L. E. HACKMAN, Y. SAHAI, W. A. T. CLARK and J. V. WOOD, in 'Rapidly Solidified Alloys and Their Mechanical and Magnetic Properties', eds. B. C. Giessen, D. E. Polk and A. I. Taub, Materials Research Society, Pittsburgh, Pa., 1986, pp. 23-26.

26. T. Z. KATTAMIS, in 'Lasers in Metallurgy', eds. K. Mukherjee and J. Mazumdar, The Met. Soc. of AIME, Warrendale, Pa., 1981, pp. 1-10.

27. R. K. H. KALIMULLIN and YU. YA. KOZHEVNIKOV, Metal Sci. Heat Treatment, 1985, 27 (3-4), 272-274.

28. M. HANSEN, Constitution of Binary Alloys, McGraw Hill, New York, 1958 and supplements by R. P. Elliott (1965) and F. A. Shunk (1969).

29. W. G. MOFFATT, The Handbook of Binary Phase Diagrams, Genium Publ. Corp., Schenectady, New York, 1978 and updates.

30. Bull. Alloy Phase Diagrams, 1980, 1, 108-9; 1982, 3, 60-74; 1984, 5, 23-30, 36-48, 348-374, 454-476, 579-592; 1985, 6, 37-42, 59-66, 149-167, 235-250 and to appear.

31. H. W. KING, J. Mater. Sci., 1966, 1, 79-90; Bull. Alloy Phase Diagrams, 1982, 2 (3), 402-403.

32. H. JONES, in 'Rapidly Solidified Metastable Materials', eds. B. H. Kear and B. C. Giessen, Elsevier North Holland, New York, 1984, pp. 303-315.

33. F. E. HAUSER, P. R. LANDON and J. E. DORN, Trans. Amer. Soc. Metals, 1958, 50, 856-

34. D. HARDIE and R. N. PARKINS, Phil. Mag., 1959, 4, 815-825.

35. A. F. BEYANIN et al, Tr. Moskov. Inst. Toukoi Khim. Tekhnol., 1974, $\underline{4}$ (1), 3-7.

36. P. RAMACHANDRARAO and G. V. S. SASTRY, Pramana,J. Phys., 1985, $\underline{25}$, L225-L230.

37. F. HEHMANN et al, Proc. P/M86, Dusseldorf, 1986, pp. 1001-1008.

38. A. CALKA, unpublished preprint submitted to RQ5, Warzburg, September 1984.

39. R. RAY, Metal Progress, 1982, $\underline{121}$ (7), 29-31.

40. P. SKJERPE, Mater. Sci. Technol., 1985, $\underline{1}$, 316-320.

41. T. Z. KATTAMIS, U. T. HOLMBERG and M. C. FLEMINGS, J. Inst. Met., 1967, $\underline{95}$, 343-347.

42. L. A. CARAPELLA, Metal Progress, 1945, $\underline{48}$, 297-307.

43. G. S. FOERSTER, U. S. Patent No. 3182390 (1965).

44. G. S. FOERSTER, Met. Eng. Quart., 1972, $\underline{12}$ (1), 22-27.

45. P. J. MESCHTER, R. J. LEDERICH and J. E. O'NEAL, to be published in Met. Trans.

46. T. E. LEONTIS and R. S. BUSK, U. S. Patent No. 2659131 (1953) and British Patent No. 690783 (1953).

47. G. S. FOERSTER, U. S. Patent No. 3219490 (1965).

48. F. HEHMANN, F. SOMMER and H. JONES, presented at American Society of Metals 'Materials Week '86', Buena Vista, Florida, 6th to 9th October 1986 at symposium on 'Enhanced Properties in Structural Metals by Rapid Solidification'.

49. G. OMORI, S. MATSUO and H. ASADA, Trans. Jap. Inst. Metals, 1975, $\underline{16}$, 247-255; G. OMORI, Trans. Nat. Res. Inst. Metals, 1979, $\underline{21}$, 192-205.

50. L. L. ROKHLIN, Russ. Met., 1979 No. 4, pp. 165-167; Phys. Met. Metallogr., 1982, $\underline{54}$ (2), 96-100.

TABLE 1 - Examples of application of techniques of rapid solidification from the melt to magnesium alloys and their effects. For a more complete survey see [20]. PW = present work, UP = unpublished.

Technique	Product	Alloy	Effect	Ref.
(1) Spray or droplet				
Gas atomization	Powder	Representative engng. alloys	Improved mechanical properties in extrusions	3-5
Rotary (spin-disc) atomization	Powder/Shot	ZK60B, ZE62	Increased compressive yield strength	6-9
Rotating electrode	Shot	ZK60A	Increased tensile and impact strength	10
Gun quench	Splat	Mg-12 to 23Al*	Solid solubility extension	11
		Mg-14 to 18Sn*		13
		Mg-16 to 23Pb*	New fcc phase	14
Catapult quench	Splat	Mg-1 to 6Mn* / Mg-0.4 to 1.5Zr*	Solid solubility extension	12
Rotating vane	Splat	Mg-8 to 25at%Ni / Mg-9 to 42at%Cu	Metallic glass formation	15-18
Two-piston	Splat	Representative engng. and novel alloys	Microstructural refinement Enhanced hardening	20,PW and UP
(2) Continuous chill casting				
Chill-block melt-spinning	Ribbon or Strip	Mg-30at%Zn / Mg-Al-Zn plus Si/Mn or Si/RE	Metallic glass formation / Increased strength and corrosion resistance	19 / 21-23
Twin-roll quench	Flake	Mg-9Li† plus Si or Ce	Increased strength at elevated T	24
Melt-overflow	Strip	AM60	Cost lower than rolled Mg foils	25
(3) Melt-in-situ				
Laser or electron beam surface melting	Treated Surface	ZK60 / MA21	Microstructural change / Improved corrosion resistance	26 / 27

* at.%   † wt.%

TABLE 2 – Elements exhibiting maximum equilibrium solid solubility C_{max}^{eq} in magnesium amounting to > 1at%. C_{max}^{eq} from [28-30]. Size factor from atomic radii given by King [31].

Atomic Number	Chemical Symbol	C_{max}^{eq} at%	Size Factor* %		Atomic Number	Chemical Symbol	C_{max}^{eq} at%	Size Factor* %
3	Li	17	– 2.4		62	Sm	∼1	+12.7
12	Mg	100	0		65	Tb	4.6	+11.4
13	Al	11.8	-10.6		66	Dy	6	+10.7
21	Sc	15	+ 2.4		67	Ho	5.4	+10.3
25	Mn	1	-19.2		68	Er	6.9	+ 9.7
30	Zn	3.3	-13.2		69	Tm	6.3	+ 8.9
31	Ga	3.1	– 5.5		70	Yb	1.2	+21.1
39	Y	3.8	+12.5		71	Lu	8.4	+ 8.3
40	Zr	1.0	+ 0.06		81	Tl	15.4	– 7.2
47	Ag	3.8	– 9.8		82	Pb	7.75	+ 9.3
48	Cd	100	– 2.4		83	Bi	1.12	+15.1
49	In	19.4	+ 4.1		94	Pu	3.2	– 4.9
50	Sn	3.35	+ 5.2					

* $100(r_2-r_1)/r_1$ where r_1 and r_2 are atomic radii of magnesium and the other element respectively.

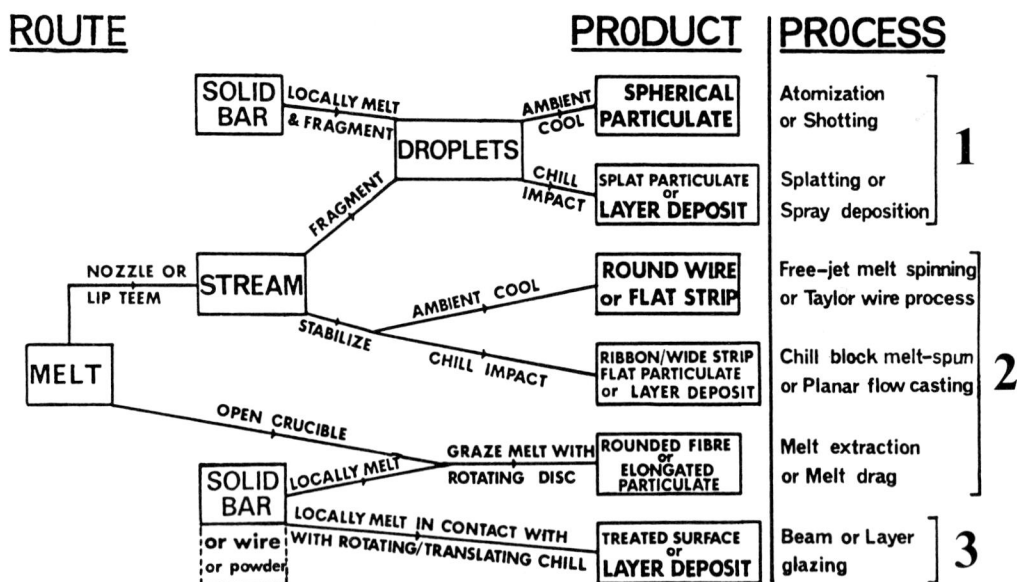

Figure 1.   Some production routes for rapid solidification from the melt and their products [1,2]. Key to methods: (1) Spray and droplet (2) Continuous casting (3) Melt-in-situ.

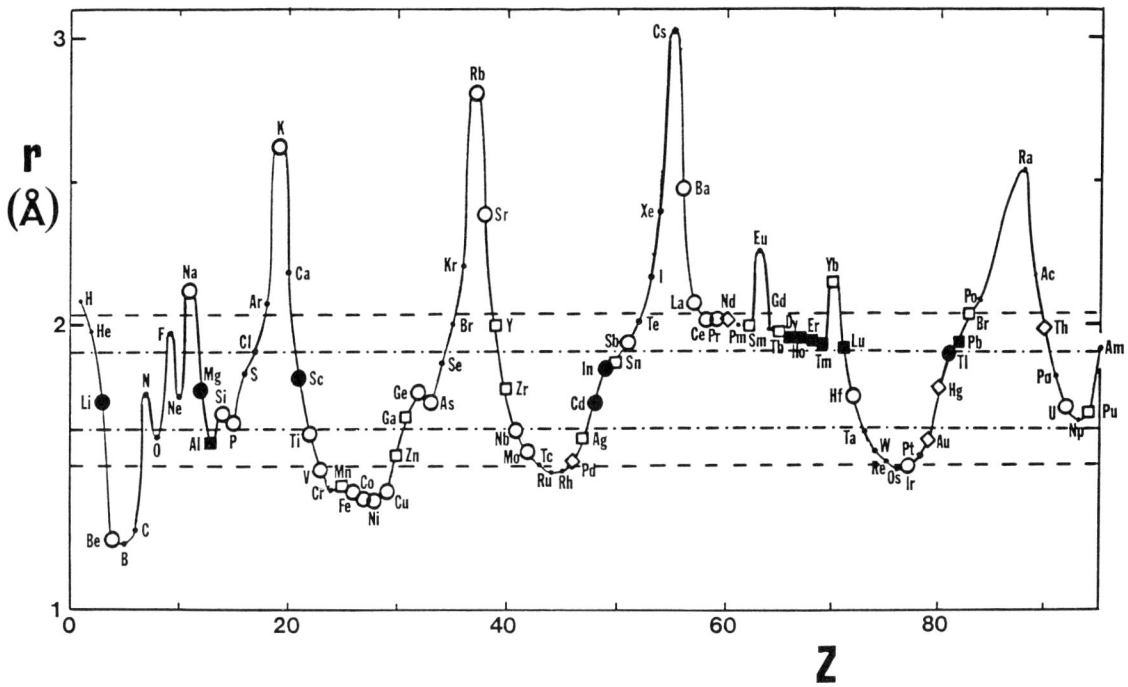

Figure 2. Atomic radius r as a function of atomic
number z compared with observed maximum
equilibrium solid solubility C_{max}^{eq} in
magnesium.

Key to C_{max}^{eq} : ● 100 to 15
■ 15 to 5 □ 5 to 1 ◇ 1 to 0.1 ○ <
0.1at% in Mg. Key to size factor
limits in Mg: – – – – $\pm 15\%$ –·–·–·–
$\pm 7\frac{1}{2}\%$.

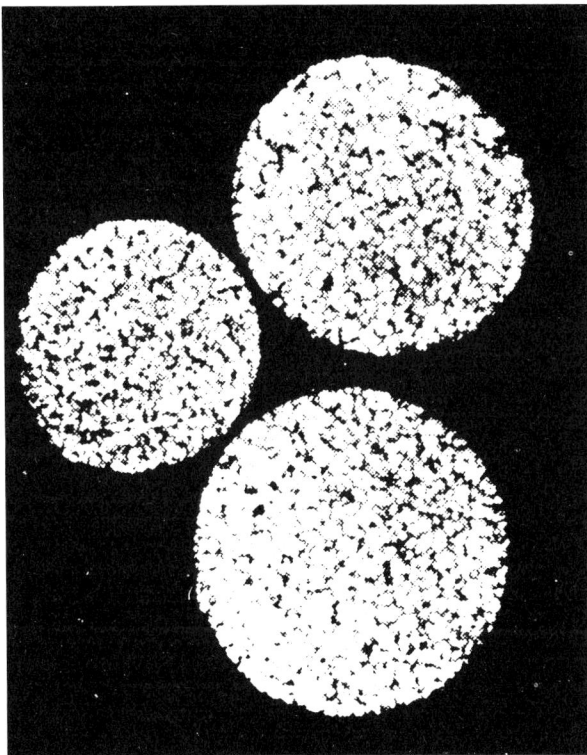

a   50μm

b   50μm

Figure 3. Microstructure of ZK60 alloy (a)
pellets made by rotating disc atomization (b) ingot cast by the direct chill
process [6].

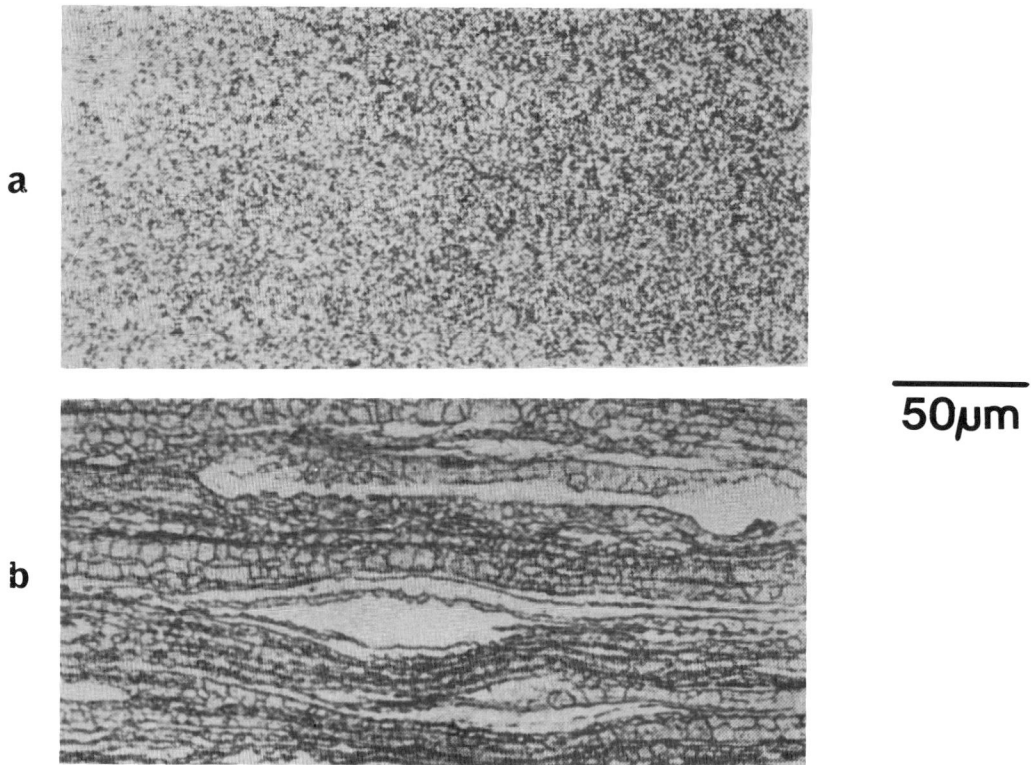

Figure 4.   Microstructure of extrusions of ZK60B
alloy made from (a) rotating-disc
atomized pellets (b) direct-chill cast
ingot material [6].

These floor beams of transport plane
get lightweight strength from pelletized
magnesium.

Figure 5.   Floor beams of C133 transport aircraft
extruded from rotary-atomized pellets
of ZK60B alloy [9].

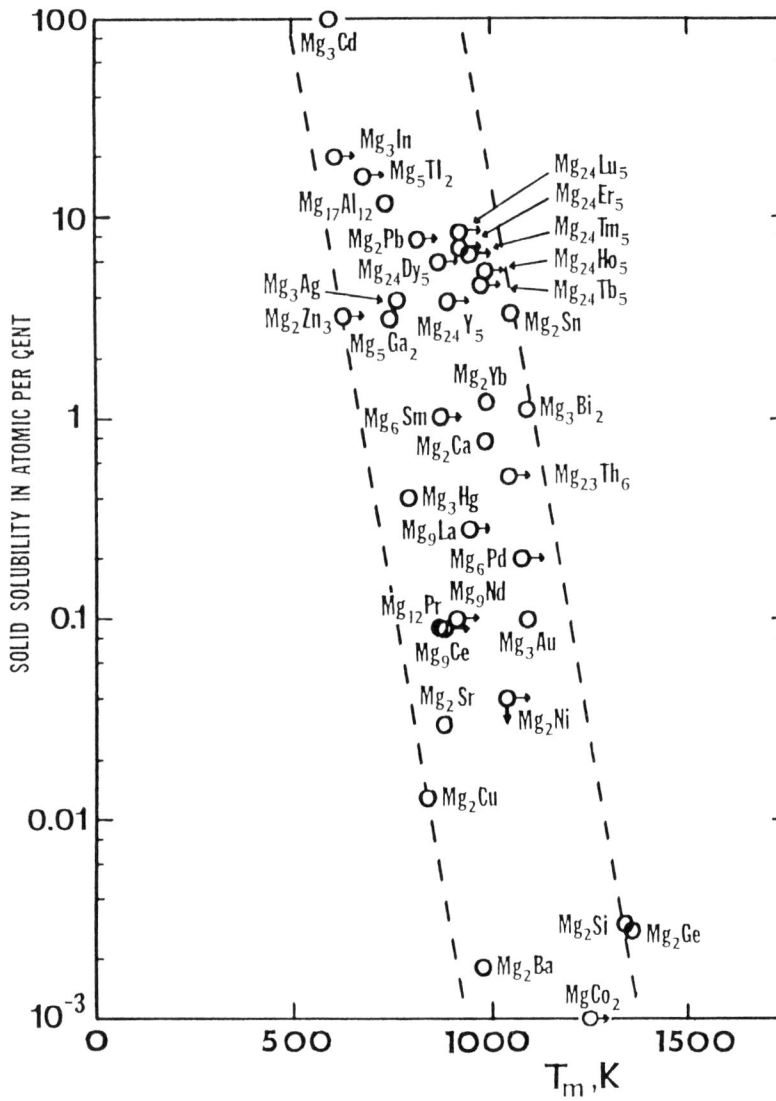

Figure 6. Maximum equilibrium solid solubility
in magnesium as a function of melting
point T_m of most Mg-rich binary intermediate phase [20], extending the plot
first given by Carapella [42].

Figure 7.

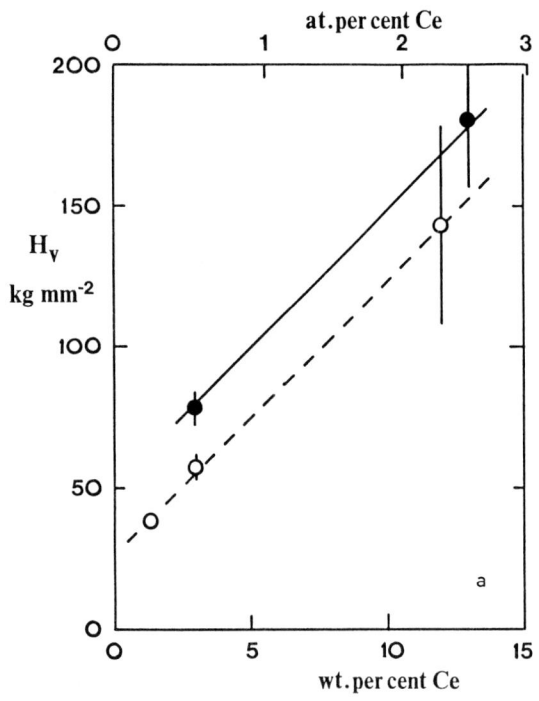

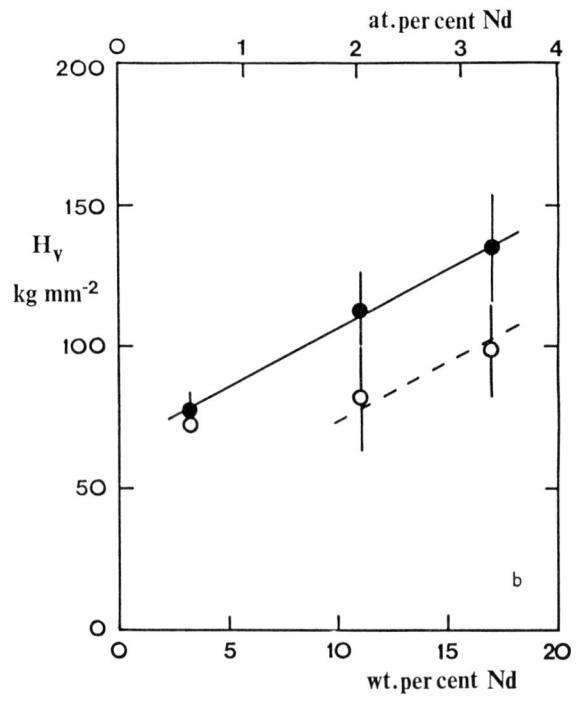

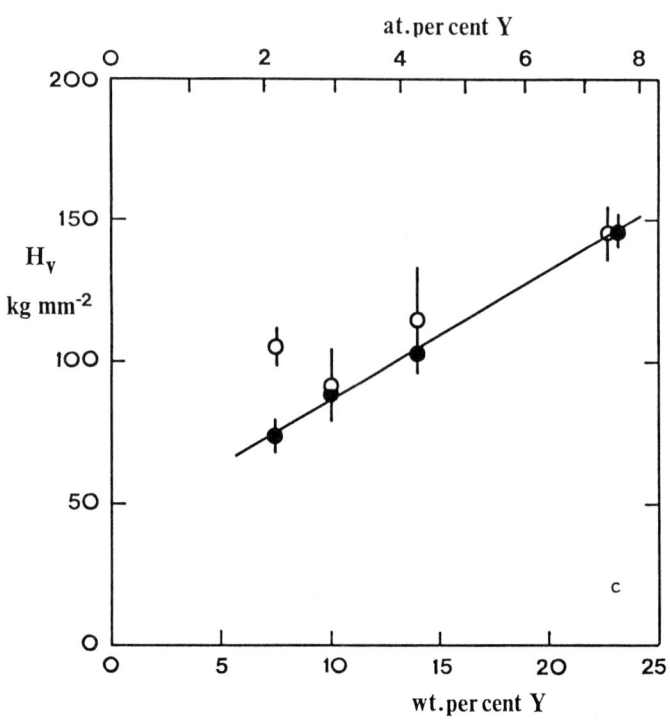

Effect of alloy content on the micro-hardness H_v of (a) Mg-Ce (b) Mg-Nd and (c) Mg-Y alloys. Closed points indicate rapidly-solidified by the two-piston technique. Open points are for chill-cast material. The point for 1.3wt%Ce is from ref. 49.

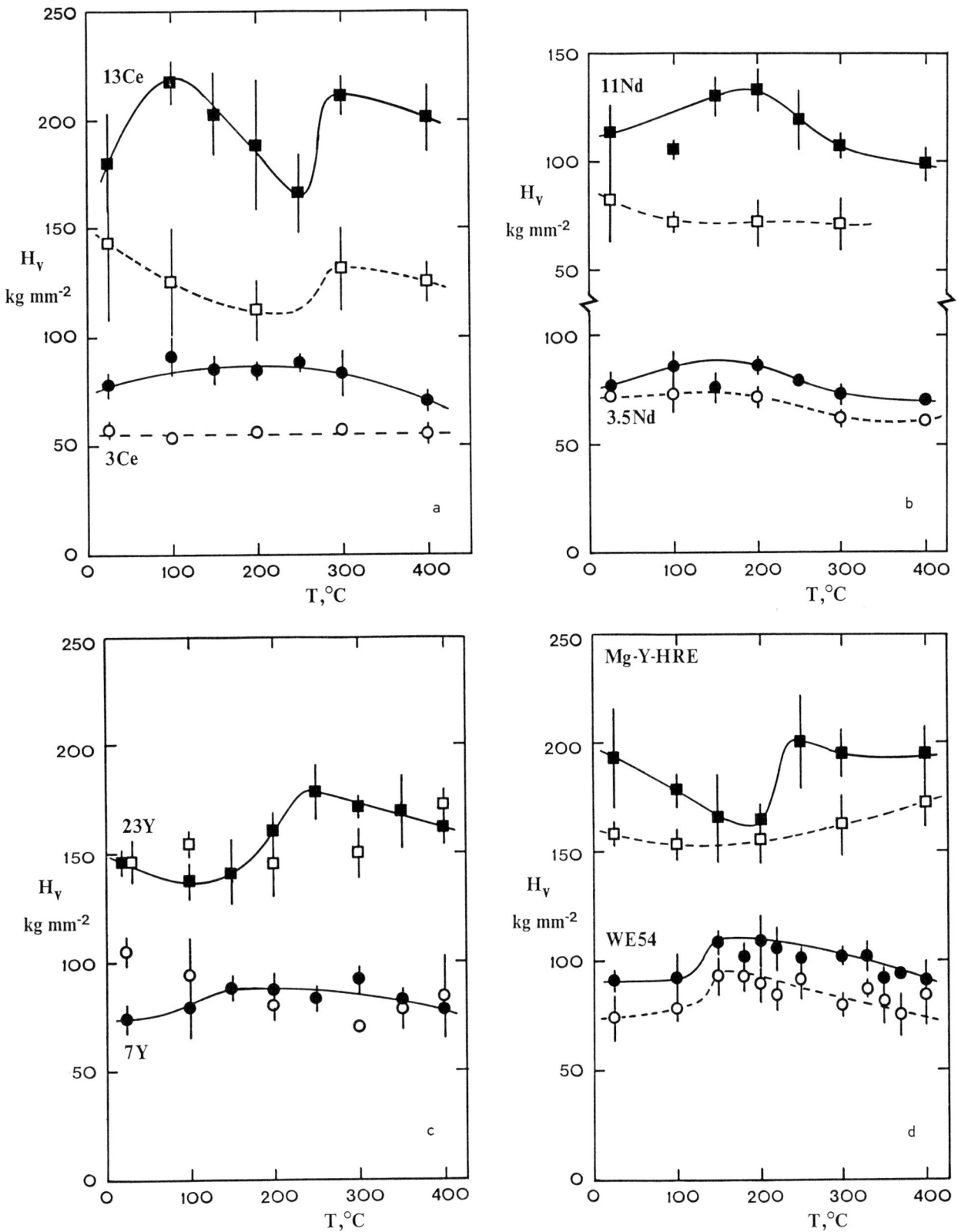

Figure 8. Effect of 1h-treatment at temperature T up to 400°C on microhardness of (a) Mg-3 and 13wt%Ce (b) Mg-3.5 and 11wt% Nd (c) Mg-7 and 23wt%Y and (d) WE54 and Mg-18wt%Y-6wt%HRE. In each case closed points indicate rapidly-solidified condition and open points chill-cast condition, with the more dilute alloy indicated as circles and the more concentrated one as squares.

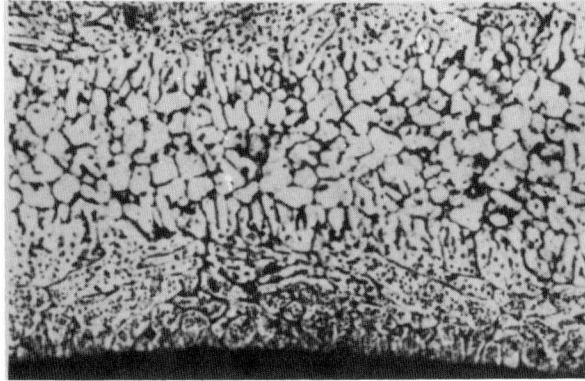

a

50μm

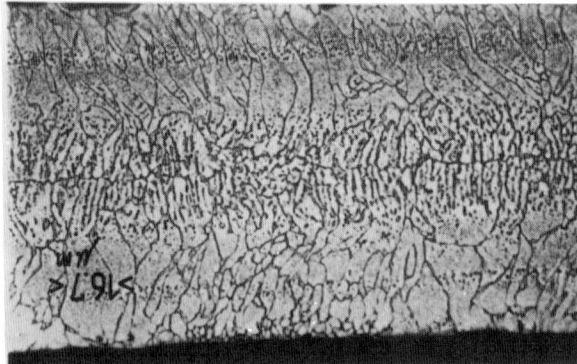

b

50μm

Figure 9. Showing essentially intact as-soli-
dified microstructure in rapidly-
solidified splats after treatment at
elevated temperature (a) Mg-3wt%Ce
after 1h at 300°C   and (b)
Mg-7.5wt%Y after 1h at 200°C.

Energy storage applications of magnesium

B VIGEHOLM

The author is in the Metallurgy Department of Risø National Laboratory, Roskilde, Denmark.

Synopsis

This paper outlines the concept of energy storage in metal hydride with emphasis on the magnesium/magnesium hydride system. An attempt is made to survey the magnesium and magnesium alloy hydride application as it may be inferred from operating, planed and intended programmes mainly involving other hydrides. The overview encloses hydrogen storage, heat storage, heat pumps, hydrogen recovery and refining.

Introduction

A number of metals and intermetallic compounds are known to react with hydrogen to form the chemical compounds, metal hydrides. This reaction can be used for energy storage if the bound hydrogen can be released easily from the hydride, i. e. if the reversible reaction

metal + hydrogen <--> metal hydride + heat

has its equilibrium at manageable temperatures and hydrogen pressures, or put in a different way, if the heat of hydride formation is moderate, (smaller than approximately 80 kJ/mole). When such a metal is exposed to hydrogen at the appropriate pressure (depending on the temperature) a spontaneous reaction will take place to form a metal hydride. This exothermal process is reversed by lowering the hydrogen pressure at the given temperature, or by increasing the temperature at the given pressure.

A massive research in metal hydrides was initiated in the fifties by the nuclear world (reflectors, moderators etc), but it was not until the oil shortage in 1973/4 that their potential as an energy medium came into focus, even then promoted as much by concern about exhaust pollution as by demands for alternative energy vectors.

In the early days of the search for hydrides with good energy storage properties magnesium was investigated in several places, e.g. (1-2) because of the high weight concentration of hydrogen, see table 1, the abundancy and relatively low price.

Table 1. Some combustion energy densities

Fuel material	MJ/kg	MJ/l	( )*
Hydrogen in MgH_2	10.9	15.7	(12)
Hydrogen in $FeTiH_{1.95}$	2.7	18.0	(14)
Heavy fuel oil	40.6	38.6	
Light fuel oil, gasoline	43.5	36.5	
Coal, medium quality	27	35	(20)
Wood	12.5	8.4	(5)

( )*: approx. value with normal packing density

Despite some positive results (1) magnesium however was soon disregarded - at least as a pure element,- because of alleged insurmountable difficulties in activating the hydrogen/magnesium reaction, obtaining sufficiently fast kinetics and preventing deteriorating contamination.

Development of magnesium alloys and intermetallic compounds like Mg_2Ni was continued, e.g. (3-4) but it was not until the end of the seventies that the properties of pure magnesium were reviewed and the mentioned difficulties largely overcome during the succeeding years (5-10).

Magnesium and the requirements of a metal hydride-based energy storage system

Unfortunately there is much more to a good storage system than high energy density. The most essential requirements are in an arbitrary order:

1 high hydrogen density,
2 good kinetics,
3 low heat of hydride formation,
4 reliability,
5 safety,
6 uncomplicated technology,
7 absence of pollution,
8 low cost,
9 availability.

These requirements - where relevant - have been investigated over the last ten years - or a little less in the case of magnesium. The status concerning magnesium may for each point be summarized to the following:

1 the highest available in an "artificial" system,

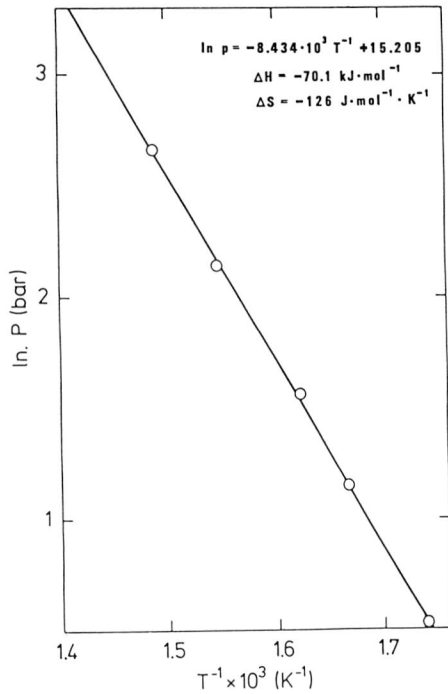

Fig. 1. Relation between temperature and hydrogen
pressure at equilibrium for the reaction
of hydrogen with magnesium.

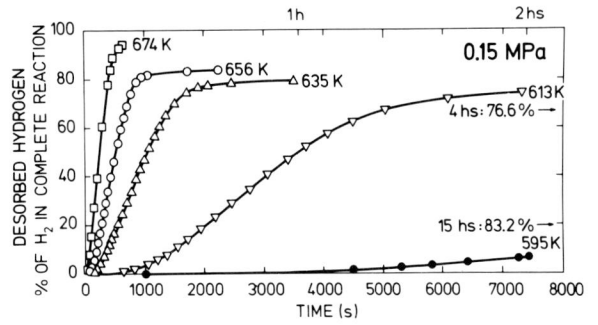

Fig. 3. Hydrogen desorption from magnesium. Same
material as in fig. 2.

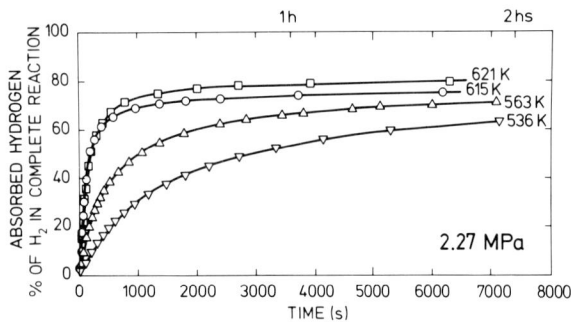

Fig. 2. Hydrogen absorption in magnesium powder
with 50 μm mean particle diameter.

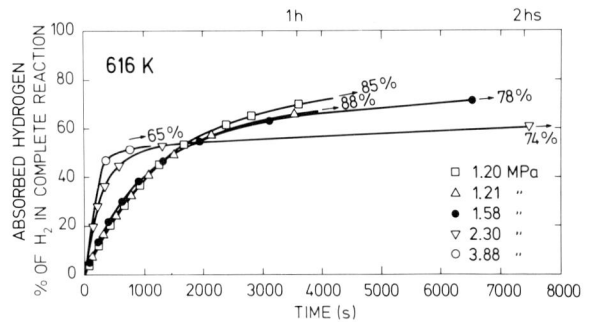

Fig. 4. Effect of hydrogen pressure on the absorption in magnesium powder at 616 K. The
ultimately recorded absorption is indicated for each pressure.

2 sufficient for most purposes but require large specific surface areas and high temperatures, >250°C for storage, >350°C for release,

3 the heat of formation is approaching the upper limit for practical applicability, thus constituting the only serious disadvantage of the element,

4 contamination may cause problems but can be avoided by simple measures,

5 specific procedures must be followed but in general the risks are similar to those of any other combustible,

6 the technology is known, but neither the heat-pump-like operation nor the containment of hydrogen are particularly simple,

7 few systems, if any, could be more satisfactory, as no polluting combustion products are formed and the only waste, MgO eventually produced by oxidation, may be considered a rather inactive and harmless material,

8 the cost of storage may be small compared to primary energy price but as it is additional, it should be compensated by reduced pollution, reduced power capacity or improved availability all of which are seldom evaluated,

9 good, unlimited resources and a manufacturing capacity which may be assumed to expand with demands.

Although the enthalpy of the hydrogen reaction has the high value of -70 kJ/mole, this represents only about 30% of the combustion energy of the absorbed hydrogen. From the comparison with conventional fuels shown in Table 1, it is apparent that thermodynamically magnesium hydride is quite an attractive energy storage medium. The high efficiency of the hydrogen combustion and the absence of polluting products - even the amount of nitrogen oxides may be reduced to insignificance - further add to the attractiveness.

The high operating temperature, the main obstacle to wide application, has been tackled by alloying. With small additions e. g. rare earths, the kinetics are improved and the practical temperature limit thereby lowered a little. Larger changes in the thermodynamic properties require substantial additions with consequent loss of storage capacity and usually increased materials price.

The magnesium/magnesium hydride system

The reaction of hydrogen with magnesium metal is a simple exothermal process:

$$Mg + H_2 \longleftrightarrow MgH_2 + 70 \text{ kJ/mole}$$

with an equilibrium hydrogen pressure of 0.1 MPa at 280°C. The temperature/pressure relation in the range 300°C to 420°C is presented in Fig. 1. The reaction rate is relatively small requiring large surface areas for practical purposes. Depending on the initial morphology, particles smaller than approximately 75 µm are required. The kinetics of the hydride formation in such particles are shown in Fig. 2, for temperatures from 536 K to 621 K. At higher temperature the reaction rate is further increased but not much, e.g. at 673 K the reaction is completed in 7-8 minutes.

Although no hysteresis is observed in the absorption/desorption cycle for the magnesium system as opposed to other hydride forming materials, the kinetics of the desorption differ substantially from those of the absorption, Fig. 3. This difference has several reasons. The most important is probably the high enthalpy, which leads to local heating during absorption and cooling during desorption, both leading to uncertainties in assigning the proper temperature to the processes.

The hydrogen absorption in magnesium follows a simple nucleation and growth mechanism. Nuclei of hydride are formed at the surface and grow more or less isotropically into the bulk by fast hydrogen transport along the metal/hydride interface. This growth continues until the hydride islands are coalesced into a continous shell, after which no further transport of hydrogen to the interior is possible.

The described mechanism has the perhaps surprising consequence, that the higher the nucleation rate the less hydride formed. The effect is seen from Fig. 4, where the hydrogen pressure is the nucleation rate controlling parameter. The nucleation and growth is visualized in the micrograph Fig. 5. The desorption follows the same pattern, only the geometry is more favourable, leading to a complete release of the hydrogen.

In most of the metal hydride systems the lattice expansion caused by the high density of hydrogen will lead to a disintegration of the material eventually forming a dust of submicron particles. This in principle occurs in magnesium as well, but with the high operating temperature compared to the melting point a spontaneous sintering takes place resulting in a coherent spongy mass, which changes little with further cycling.

The mechanical deformation caused by absorption/desorption cycling leaves a highly distorted material with many microcracks. This explains why the described nucleation and growth mechanism does not result in a very low absorption. Although the surface is activated by the cycling and the nucleation rate consequently increased, the effect of the increased specific surface is dominating. With particle sizes in the range 30 - 75 µm the degree of reaction reaches 85 - 90%, at which level it may stay for thousands of cycles in pure hydrogen. With initial particle size smaller than 20 µm a 100% reaction may be approached and maintained, but as the handling becomes more hazardous, this may not be practical.

Energy related applications of metal hydrides

Although the enthusiasm for metal hydrides as energy storages diminished with the falling energy price from about 1980, much of the effort invested in basic research and application technology in the seventies has been followed up. In the last few years a variety of metal hydride based equipment has been marketed and energy related applications demonstrated. Most of the implementations involve relatively expensive low temperature hydride systems partly because of the convenience of room temperature operation, partly because the applicability of magnesium based hydrides - as mentioned above - is only recently appreciated.

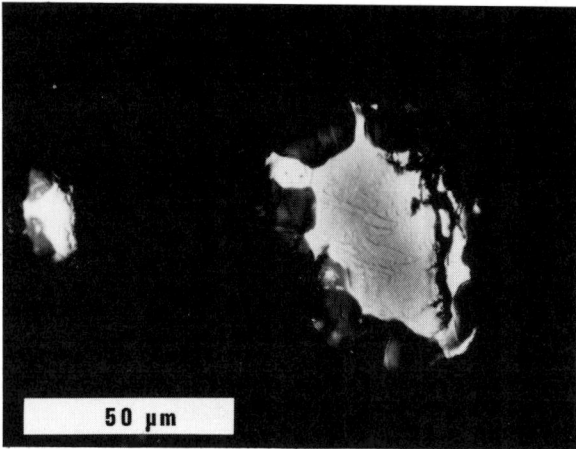

Fig. 5. Hydride formed at the surface of magnesium particles at 2.8 MPa and 400°C

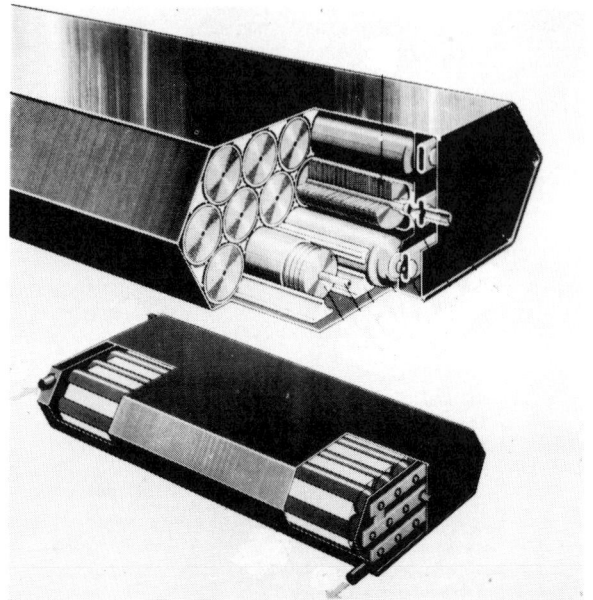

Fig. 6. Commercial hydride fuel tank (Mannesmann, HWT AG).

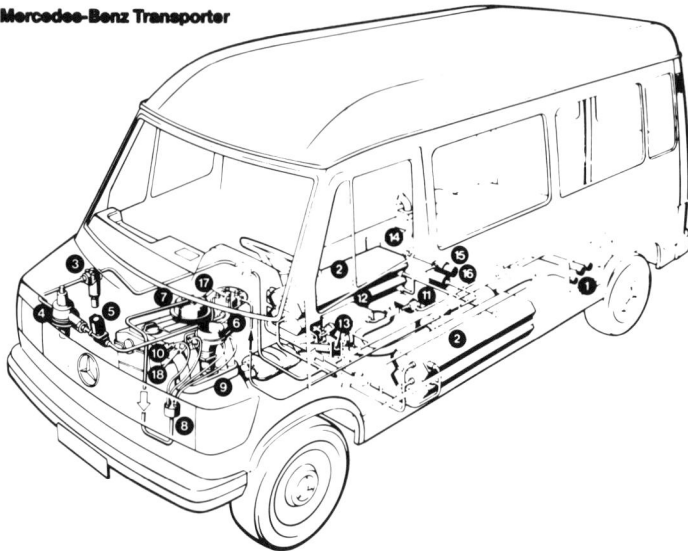

Mercedes-Benz Transporter

Fig. 7. Hydride fueled Mercedes-Benz van with four hydride tanks, (marked 2).

Fig. 8. Hydrogen storage plant for 2000 Nm³ at max pressure 5 MPa.

The working principle is the same for all metal hydrides, so although actual operating conditions may favour one particular material it is to be expected, that at least in some applications an economy-induced switch to magnesium will be seen.

The energy related applications of metal hydrides can be divided into:
                    hydrogen storage,
                    heat storage,
                    heat pumping.
Other usages less directly related to energy are:
                    hydrogen recovery,
                    hydrogen purification,
                    hydrogen compression,
                    thermal sensing and control,
                    barometric sensing and control,
                    nuclear applications

Hydrogen storage.

An important feature of hydride storage is the relative insensitivity to plant size. This lies behind the most spectacular implementation seen so far, as fuel supply for landbased vehicles. Fig. 6 shows the hydride fuel tank (Mannesmann) used by Daimler-Benz in their latest hydrogen fuelled vehicle tests. An "x-ray" picture of the hydride fuelled Mercedes-Benz van "Transporter" is shown in Fig. 7. The van has a range of 120 km (urban traffic conditions) with a top speed of 130 km/h for one charge of the 560 kg hydride tank system. The hydride material is a complex, low temperature alloy developed over several years for this specific purpose. It seems likely that magnesium being far less expensive and having at least four times the hydrogen/weight concentration may be a better choice for part of the tank system. The only real problem with magnesium is, that with improved combustion efficiency the waste heat available at the required temperatures ($>350^{\circ}$C for Mg, $>200^{\circ}$C for Mg-alloys) is insufficient for the release of hydrogen. Present efforts to develop higher temperature combustion chambers may change the situation, but magnesium hydride can never stand alone as a traction fuel. Daimler-Benz has an impressive record of hundreds of thousand miles testing of private cars, vans and small buses, also with combined hydrogen-petrol systems, but there are several other manufacturers, who have gained practical experience with hydride fuelling.

One particular traction application of the hydride fuelling should be mentioned, namely powering of mining vehicles. Hydrides here are very attractive alternatives to electric batteries. The idea was conceived at Denver Research Institute, where the first prototypes were also developed. A practical test programme was commenced a few years ago.

Performance and economy of stationary large scale load levelling plants have been analysed (11) but so far no plant has been constructed. Small plants suitable for storage of solar, wind- and other renewable - and variable - energies in stand-alone systems are currently being investigated in many places. Implementations may be expected in the near future.

Heat storage.

In a closed system heat may be stored by decomposition of a hydride. When the system is removed from the heat source and the temperature reduced the re-formation of the hydride will release the stored heat. By controlling the hydrogen pressure and the heat transmission the temperature may be controlled to some extent. In principle there is no difference between hydrogen storage and heat storage provided gaseous hydrogen can be contained. The 2000 Nm^3 Mannesmann plant shown on Fig. 8 is going to be used for waste heat utilisation. It contains low temperature hydride and the heat capacity can be estimated to 2 - 2.5 GJ or the equivalent of the heat from combustion of 50-70 l oil. Commercially only magnesium with the high energy of hydride formation seems to have any future, but probably not the brightest as high temperature waste heat is found in a limited number of processes like metal casting only.

Heat pumping.

Metal hydride heat pumping can be performed in closed systems in two ways, one requiring supply of mechanical energy, one all thermal. In the first case two identical containers, one holding a hydride, the other the corresponding metal are connected through a gas compressor. Heat from a low temperature source is absorbed by dissociation of the hydride. The released hydrogen is mechanically compressed and fed into the second container where it reacts to form the original hydride. The heat of reaction is the same in the desorption and the absorption, but as the latter takes place at a higher pressure, the heat is given off at a higher temperature. When the process is completed the direction of flow is reversed and a new cycle can start.

The second type of pumping requires two hydrides with different thermodynamic properties. The pumping action may be explained as follows. The higher temperature metal/hydride system is termed A, the lower temperature one B. High temperature (waste) heat is supplied for desorption of hydride A. The hydrogen is led to system B to form hydride at an intermediate temperature. This reaction continues until all the hydride, A and all the metal, B has reacted. The high temperature supply is now cut off and metal, A allowed to cool to an intermediate temperature, and B to a low (ambient) temperature. The ambient heat is now desorbing system B and the hydrogen used for forming hydride A giving off heat at the intermediate temperature. Waste heat is again supplied, and A will rise to the high temperature. It can be shown that the pumping efficiency of this process approaches the Carnot-efficiency. Overall performances are proven comparable with those of conventional heat pumps but the hydride systems have the advantage of being less dependent on the operating temperatures as the controlling factors are mainly the enthalpies of hydride formation.

Complex systems comprising three or more hydrides, and systems using partly thermal partly mechanical energy have also been devised. It looks as if some of the systems are close to commercialisation. Recently a noticeable interest in high temperature heat pumps (e.g. for geothermal heat extraction) has been shown in Japan and in Europe. This could develop into a potential area for application of magnesium.

Other applications.

Hydrogen recovery. A large hydride plant for recovery of boil-offs from the LH$_2$ supplies at Cape Kennedy was at an engineering stage when the

"Challenger" accident took place. Undoubtedly the plant will be built, when the space programmes are resumed. A plant for recovery of waste hydrogen from ammonia production was operating for 8 to 9 months apparently successfully. Both plants were devised by Ergenics, N.J.

Hydrogen purification may be carried out very efficiently by absorption in a metal. The impurities may either form stable compounds or stay in gaseous form. In both cases it will be possible to remove them, before the hydrogen is desorbed from the hydride. Purification units in many sizes are commercially available. The present units are all based on low temperature hydrides.

Hydrogen compression. By application of several different hydrides in a way somewhat similar to the heat pumping, chemical compression to over 50 MPa is possible. A range of such compressors, which additionally act as purifiers, are marketed in the U.S.A. (e.g. Ergenics) and in Europe (e.g. HWT AG). The cost of compression is similar to that of mechanical methods, but the hydrides have a number of advantages such as reliability, safety and low noise levels.

Sensors. Thermostatic, directly operating devices can be produced with very small amounts of hydride. A single gramme is sufficient to operate a several inch valve against high pressure. With the number of known hydrides the entire range from -20° C to $+450^{\circ}$C may be covered in small intervals.The main objections to using hydrides are that many of them show hysteresis i.e. absorption and desorption temperatures at the same pressure differ, and that they have relatively slow responses to changes in temperature or in pressure because of the reaction heat to be transferred. Possibly the most widely used fire alarm in airplanes is based on desorption of hydrogen from a hydride.

Nuclear applications. The high atomic density of hydrogen makes hydrides suitable for neutron interactions (moderators, reflectors, shields), but the main interest at the moment seems to be removal of tritium from cooling circuits.

Conclusions

Metal hydrides have been investigated thoroughly all over the world during the last 10 to 15 years. From a technological point of view they have been proven to be very functional in numerous ways. The large scale energy related applications however have not seemed sufficiently competitive to encourage much investment in the new and unfamiliar technique. Magnesium re-entered the scene a little late but is catching up and will have a role to play in any massive implementation.

The fact that many spin-off applications have been developed to a commercial stage may be very valuable, as it makes the industry and the public aquainted with the hydride concept.

The fossil energy sources have been too hard to compete with in the eighties, but the fast growing concern over pollution and ultimately the CO_2-emission may change the situation drastically in few years. The future for the hydrides may be a little uncertain, but a general departure from fossile energies is inevitable and this will lead to an enhanced demand for storages. At present there is no better solution to this than hydrides and for large scale uses no other material than magnesium.

References

1. J. F. Stampfer, C. E. Holley and T. F. Suttle, The Magnesium Hydrogen System, J. Am. Chem. Soc., 82 (1960) 3504.
2. J. J. Reilly and R. H. Wiswall, The Reaction of Hydrogen with Alloys of Magnesium and Nickel and the Formation of Mg_2NiH_4. Inorg. Chem. 7, (1968) 2254.
3. B. Tanguy, J-L.Soubeyroux, M. Pezat, J. Portier and P. Hagenmuller, Mater. Res. Bull. 11 (1976) 1441-1448.
4. D. L. Douglass, in A. F. Andresen and A. J. Maeland (eds.), Hydrides for Energy Storage, Pergamon, Oxford (1978) 151-184.
5. J. Isler, Thesis: Étude Cinetique des Reactions D'Hydruration et de Déshydruration du Magnesium et du Mélange Mg - (LaNi$_5$). L'Université de Dijon, France (1979).
6. B. Vigeholm, J. Kjøller and B. Larsen, J. Less-Common Met., 74 (1980) 341.
7. B. Vigeholm, J. Kjøller, B. Larsen and A. S. Pedersen. In: Hydrogen as an Energy Carrier. Proc. 3rd. Int. Seminar, Lyon, 25-27 May, 1983. Reidel Publ. Co., Dordrecht, (1983) 442-456.
8. A. S. Pedersen, J. Kjøller, B. Larsen and B. Vigeholm, Int. J. Hydrogen Energy, 8 (1983) 205-211.
9. B. Vigeholm, J. Kjøller, B. Larsen and A. S. Pedersen. On the Hydrogenation Mechanism in Magnesium. In: T. N. Veziroglu and J. B. Taylor (eds.), Hydrogen Energy Progress V. Proc. 5th. World Hydrogen Energy Conf. Toronto, 15-20 July 1984. Pergamon, New York (1984) 1455-1463.
10. A. S. Pedersen, J. Kjøller, B. Larsen and B. Vigeholm, Int. J. Hydrogen Energy, 10 (1985) 851-857.
11. E. Schmidt-Ihn, O. Bernauer, H. Buchner, and U. Hildebrandt, Storage on Hydrides for Electrical Load-Equilibration. Proc. Hydrogen as an Energy Vector, Comm. European Comm., Bruxelles (1980) 405 - 420.

Applications of magnesium in aerospace

G B EVANS

The author is recently retired from British Aerospace Civil Division, Hatfield, England.

SYNOPSIS

Magnesium is used both as an alloying agent in aluminium alloys for aerospace and as the base metal in alloying.
The very low density enables a widespread use in aerospace although because of its properties uses seem restricted to the more massive parts. The extent of its application is illustrated mainly from European sources, and the possibility of widening this use is discussed from different aspects.

1   INTRODUCTION

There have been several papers at recent conferences in the USA dealing with aspects of the use of MAGNESIUM in all fields, but naturally although metal production is quoted on a world-wide basis[1,2] the applications examined have been largely related to North American uses and as far as aerospace is concerned very much related to the huge military market (Fig 1).

Magnesium in structural aluminium alloys

The studies referred to show the very large background demand for magnesium as an alloying addition with aluminium alloys. Despite the escalating introduction of fibre-resin composites in main and auxiliary structures the improvements in strength and performance of this group of alloys have maintained demand in current civil and military aircraft.

This may not continue to be the case in the next generation of projects as the properties of the Al–Li alloys more nearly approach them and have a density advantage even at very high strength.

It must be remembered that the equation which determines the material selected is not simple and for a while the lower production cost and high properties of the Al–Zn alloys may hold their own.

It has much to do with the trend in fuel prices. If the oil price stays low then the Al–Li alloys have a harder job displacing the Al–Zn–Mg alloys in high-strength structural applications.[3]

Magnesium-based alloys

It is suspected that the conference organisers did not have that aspect in mind but rather the direct use of magnesium in magnesium-based alloys, even though the terms of reference given for the paper requested the widest approach in aerospace use. This also seemed to point to a closer look to establish current applications and future trends in European practice, especially as the meeting is in London. Where better to look for these than at the Farnborough 1986 SBAC Airshow? Therefore, coupled with a more detailed discussion with suppliers and constructors internationally, this paper is based on that theme.

2   APPLICATIONS

Castings

One thing was immediately apparent on walking around the exhibition: most of the parts displayed were castings of one sort or another. So also were the parts identified on the completed assemblies. The second feature was the seemingly large number of such examples despite all that has been said about the corrosion behaviour of magnesium alloys. The following examples illustrate this in more detail.

Engines

Two primary uses were identified on many engines displayed: the casing of the compressor end of the engines and massive gear box casings containing for the most part reduction gears or ancillary equipment drives. The AVCO Lycoming bypass engines which have enabled the quietest airliner in the world to be made, the BAe 146 (Fig 2) have bypass casings in magnesium (Fig 3) as did Pratt and Witney engines on display.

In correspondence AVCO have said that they are developing a new version with a stainless steel casing to improve corrosion resistance and to

MAGNESIUM CONSUMPTION 1984*

TONS.

CATEGORY	WESTERN EUROPE	WORLD TOTAL
AL. ALLOYING	35,400	118,200
DIE CASTING	18,400	30,400
STRUCTURAL PRODUCTS	700	6,600

* NEW FRONTIERS IN PRODUCTIVITY WITH MAG. DIE CASTINGS. DWAIN M. MAGERS. AFS/CMI SPECIAL CONFERENCE JUNE 1985

Figure 1   Distribution of the use of magnesium in alloying aluminium (some for aerospace) and in magnesium base alloys - in Europe

Figure 2   BAe 146

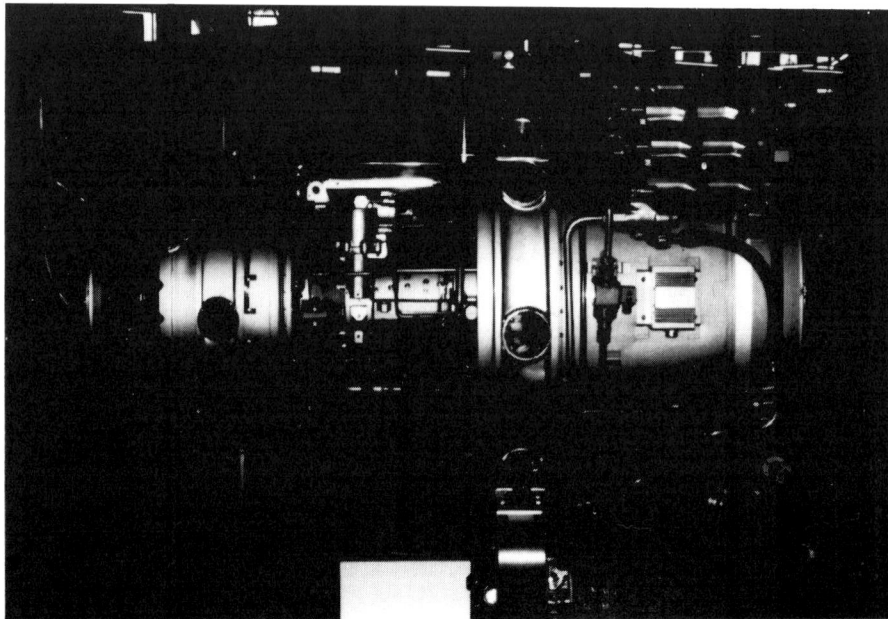

Figure 3   View of AVCO Lycoming Textron ALF 502R. Turbofan with magnesium alloy castings

enable an increase in thermal stability to occur, to allow closer blade casing tolerances and increased engine efficiency. Another engine manufacturer said that they made two versions of their engines that had dual civil and military roles. While the civil engines had magnesium casings the military versions were of cast aluminium alloy. Nevertheless engines of this sort have been in operation for many years so clearly with correct protection, and this includes the need for special attention to galvanic corrosion protection where other metals are attached to the magnesium, such a use has many advantages. The new high-temperature resistant, high-purity magnesium alloys have their place here.

The second group of very large castings that were evident were the gear casings; for example, that used by Kent Aerospace Castings on the huge Rolls Royce RB211 engine. These castings, as with others on such engines, will doubtless be protected by the exceptionally good heavy duty coating applied by Rolls Royce for many years. There are of course many smaller castings on engines of a similar nature, but a book would be needed to describe them all.

Helicopters

At one time these vehicles had magnesium alloy sheet fuselages, but today I understand this has been replaced to some extent by resin-fibre composite. Undoubtedly the vulnerability of such thin structures to corrosion has contributed to this change. Unlike the aluminium alloys, magnesium sheet has never been blessed with the equivalent of the weak but corrosion-protective pure aluminium layer without which the aluminium alloys would have similar problems.

However, very large castings in magnesium were much in evidence, examples being the gear boxes used on the engines and in particular the special boxes for the rotor heads. A very large number of the world's helicopters have this in common, as it seems will future designs with even larger castings.

Airframes

There have been many applications of magnesium castings on the airframes of fixed-wing aircraft, military and civil.
The amount varies from constructor to constructor, to some extent dependent on his individual development of surface coatings. This is not only on account of the corrosion behaviour of such parts, which should be largely overcome by the use of higher-purity alloys, but also aspects such as wear which feature in the need to use surface coatings on other metals as well.

The exceptional casting potential of the magnesium rare-earth alloys, coupled with their ease of welding, made the use of very complex canopy castings on many military aircraft their own from the time of their invention. Some of the early examples described by Evans in 1969[4] for the two-seat Vampire and Venom aircraft have hardly been surpassed (Fig 4). They gave excellent service and it is therefore not surprising that this application continues in modern projects.

The spectacular Hawk which featured in the Red Arrow display at Farnborough has cast canopies as do the Spanish CASA. These and similar examples were featured on several stands (Fig 5). Another use on many aircraft has been the pulleys for the wire flying controls. Such an application has been said by one European constructor to be not viable due to wear on the pulley, but BAe have used such pulley systems in magnesium alloys since 1960 on most civil aircraft, this being made possible by the introduction of plastic coating, in this case nylon. Evans[4] has previously described this development and the influence on improved wire life. The pulleys treated in this way did not wear, nor did foot pedals or hand controls.

Another common area where the stiffness of magnesium alloy casting has advantage is in the hand controls in the cockpit - it has been extensively used in the control columns and wheels, the central pedestals and wheels such as the trim wheels etc. The Trident and BAe 146 cockpit is an example of this. The similar use by Fokker on the new F100 follows satisfactory service on the F28 aircraft of that company. The interesting set for the Breguet Atlantic is shown in Figure 6.

The exhibition stands of Kent Aerospace and LMI France show many other casting applications including complex pipes for pressurisation systems (BAe 146, 125 and AIR BUS).
The control mounting box of the Hawk was the subject of an interesting study. This involved a comparison of different methods of making the control base and comparison of weight and cost from three different founders for the same part.

There are many fixed-wing civil aircraft with many years of successful service behind them which have up to 250 different cast magnesium alloy parts on each aircraft, used in the various ways described, despite the fact that there has been a tendency for forgings and castings to be reduced over this period because of the widespread use of cheap automatic complex machining of parts from solid.

Missiles and similar

My personal experiences in the missile field have been less than in the others and in any case people tend not to talk about this area as much. It is clear that many simple castings are used in such structures, based on magnesium alloys.

It is interesting that one large constructor said that they had used magnesium alloy parts successfully so far as the weaponry was concerned, but had found it difficult to protect the parts from corrosion during machining and work prior to final finishing and assembly.

This is an aspect which has rarely been discussed concerning any metal at the many corrosion conferences, although with steel and aluminium it features just as greatly in most aircraft constructors' workshop practices. Indeed the reject rate of metals is often high and very costly when the part is nearly finished.

It seems a pity that this should prevent magnesium from taking its place where it would otherwise have an advantage and this aspect should be

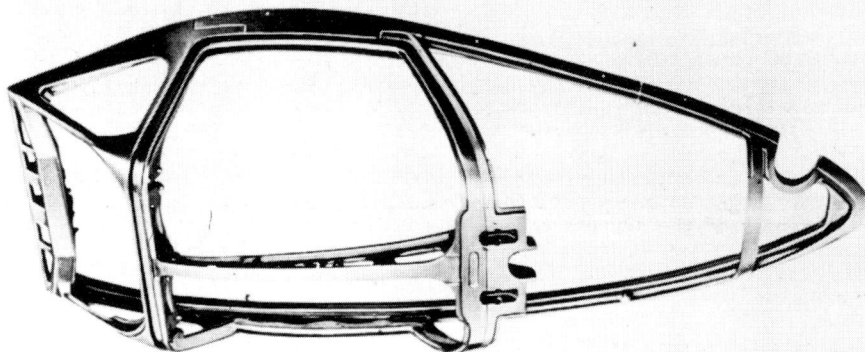

Figure 4   Side by side two seater canopy casting of the 1960s

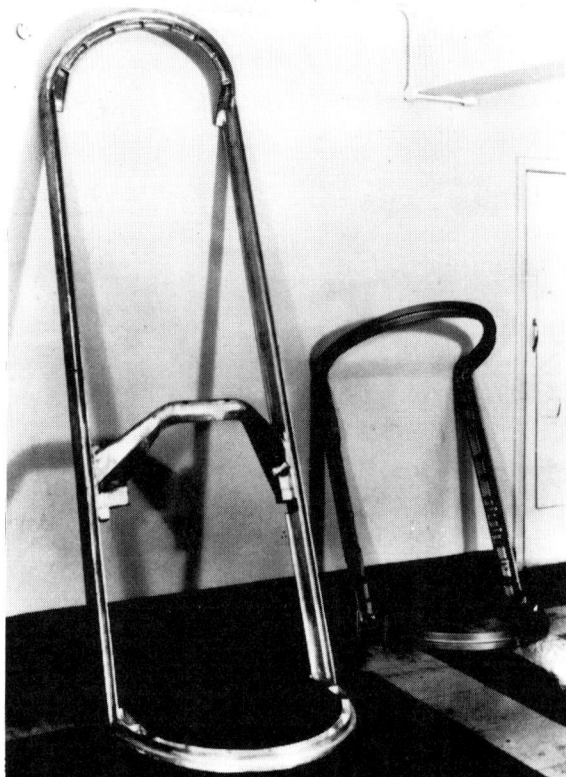

Figure 5   Example complete casting for canopy to illustrate complexity possible. The smaller canopy is for a CASA aircraft

considered. Methods are available but with all metals very good house-keeping and great vigilance is needed – magnesium is no exception. Indeed the point made in 'Corrosion and Design'[6] is that whereas surface corrosion on some of the aluminium alloys can include intercrystalline corrosion, which is difficult to inspect, this is not likely to be the case with magnesium alloy parts.

Different casting methods

The papers referred to describing North American practice included the increased use of investment casting. You will note that although shown on the stands at Farnborough 1986 the parts were small and not predominant. One might have expected that with the increased awareness of the advantage in overall cost of very close-to-form supply to the constructor, this sort of thing would have also been shown in Europe.

LMI showed parts and Industrial Precision Castings have a few parts but for a quick over-view of the several different foundry methods to achieve this objective may I refer you to Stirling Metals Ltd brochure.

3   FORGINGS

Compared with the well-known list of founders who manufacture magnesium alloy castings, fewer forgers come to mind, nor were forgings in the same evidence at Farnborough.

Indeed one North American forger said that they had not had a new forging project for 15 years. Some forgings are current in the UK, but while some of these are of substantial size, they are of course less complex than many of the casting applications.

What were thought to be the future applications for magnesium forgings and has this gone wrong?

Papers by E Stewart-Jones in 1970[7] indicate the advantages of forgings and note that magnesium may not be as easily forged as aluminium alloys but is very much more easily machined. Indeed he quotes that if malleable cast iron has an index of 1, then aluminium has a machining index of 1.5–2, but magnesium alloys 5. The concern of the day with aluminium forgings was stress corrosion from which many constructors were in some difficulty – magnesium alloys did not suffer from this.

Nevertheless the papers also show the need for quantity production even of close-to-form forgings in order that the cost may be competitive with other methods of shaping. It will be realised that NC machining had become the production byword, in the airframe industry at least, just before this time.

It could be that the cost balances have changed through the years, as has the realisation that protection is a must with all metals for the long lives required of modern aeroplanes. Similarly, again especially in Europe, the aircraft industry has changed its form and generally much longer production runs have to be negotiated for a project to be viable.

Perhaps a rethink in this area would be profitable to the designer. The simple comparison of common alloys in the two forms would suggest greater use of forgings without taking into account the developments that are the province of other papers in this conference.

4   INNOVATIVE APPLICATIONS

There have been many interesting applications which I have been privileged to assemble for you. Perhaps one of the most interesting was the Transall Air Cargo System (MILITARY). The use of magnesium chains to secure the heavy military loads must surely illustrate the almost endless purposes to which the magnesium-based alloys can be put (Figs 7,8), a fitting extension to the complex canopy casting of the 1960s with which the paper opened - these chains were also in RZ5.

5   GENERAL DISCUSSION

Several points of interest arise from this brief survey.
Constructors have the view that magnesium corrodes easily compared with other metals; I have participated in recent conferences in the USA where the military authorities have been very hard-hitting with suppliers on account of the poor service record of magnesium alloys, much related to corrosion.

This must also be true in the civil field among front-line operators, judging from the guide lines to good protection in design that IATA have produced. Bearing in mind that this has not been the reported experience with all UK aircraft it is interesting to consider some of the recent corrosion protection studies published.[9] Robinson showed that many of the methods in use had not provided good protection and said "US military specifications could be significantly improved if they were patterned on the English system", referring directly to DTD 911C.
"If an equivalent specification had been in effect in this country for military applications then the various parties, including Westinghouse, would have been spared the loss of thousands of dollars. The finish system which failed would not have been allowed ie no primer employed and a single coat finish which was highly lead-pigmented".

Indeed the military representative said at the conclusion of the meeting that the US Government would be prepared for aircraft to have magnesium alloy parts if they were treated in this way.

However, in my view this is not the only contribution to the application of magnesium alloys that should be taken into consideration:

(a)   The method given in DTD 911C and advocated by many until now is really based upon the ability of the coating to be impermeable to moisture, but does not include any QA methods that will ensure that this is the case.

In my view this philosophy will fail more readily than one based upon inhibition. This means that the primers must provide a fully available supply of inhibitor at the surface as well as, and perhaps instead of, being without porosity.

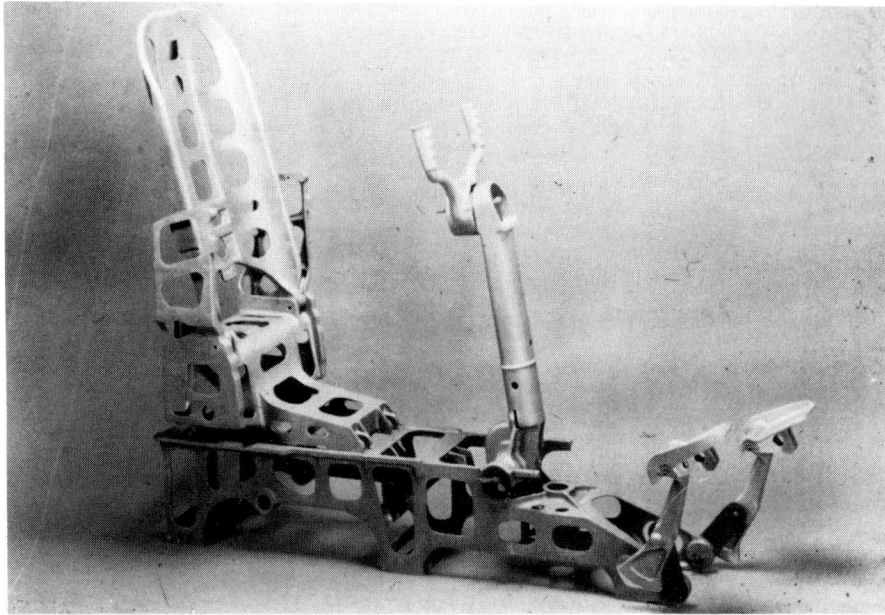

Figure 6 Pilot's seat etc for Breguet Atlantic.
(Alloys RZ5, T26 and GA9.) (LMI)

Figure 7 Interior of Transall

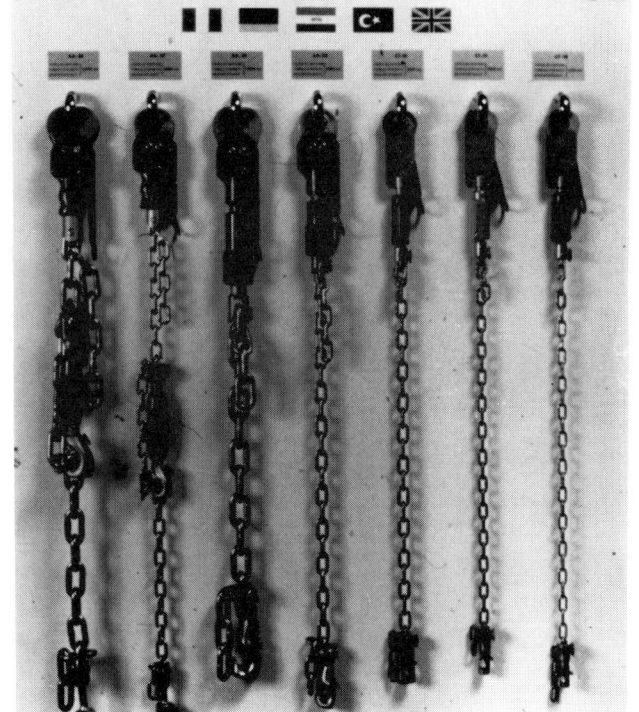

Figure 8 Securing chains, breaking load
1200–15 000 dans, for securing on board cargo.
Magnesium alloy RZ5 and H T steel

Such methods have been used in the UK with plastic coated surfaces but not with paint coatings in general. With modern chemistry economic coatings with these attributes should be possible which would be in advance even of DTD 911C.

(b)   The improvements in corrosion behaviour attributable to magnesium alloys by control of the elements iron and silicon, both in the melt and on the surface, have not been taken adequately into account. The position is therefore much as it was with the high-strength aluminium-zinc-magnesium alloys and stress corrosion until similar measures afforded similar degrees of improvement. Now these alloys are fully used, without special permission as it were, and this should be the position with magnesium alloys which include these latest developments.

Undoubtedly if this were agreed the major civil aeroplanes in which many of the applications contained in this paper are resisted could benefit from the wider use of magnesium alloys.

(c)   In addition to the guidance of DTD 911C, the British designs have apparently also benefited, by emphasis on the importance of full galvanic protection with its even greater protection than the full sealing of joints recommended in UK military and civil design procedures.

It is of course true that if the joints are fully sealed then bimetallic corrosion can be restrained, just as with coatings only more so, but no method of inspection is available which will guarantee that this has been achieved. All those who have designed and built integral fuel tanks will know the truth of this.

(d)   The reason given by the forging fraternity for the reduced use of forgings - the corrosion potential of magnesium alloys - would therefore seem questionable also. Do forgings corrode more easily than castings? Surely not. The risk of contamination with iron on the surface may be greater during production but this should be easily avoided by available procedures. It is felt that other reasons contribute. This might be the dramatic change of cost related to numbers ordered, so that in design offices magnesium forgings have gone out of fashion. The longer production runs of projects and the advances discussed should at least make it necessary for the designers to look again.

New uses

Are there new uses for magnesium alloys to the advantage of the aeroplane etc? There is every reason to believe that there are complex parts, stiffness critical, where the above improvements would be to the advantage of the project. Should there be more?

The US foundry industry has worked on premium quality castings, thin-walled and large, for complete components such as leading edges, elevators etc. This has been in both aluminium and magnesium.

An important aspect of parts such as this is that they are relatively easily replaced in the event of unserviceability for whatever reason.

With the magnesium alloys correctly selected, such purposes would have the added advantage that they could be repaired quickly by fusion welding. This attribute has its place in keeping the first cost of production errors down at any stage - a foundry problem, incorrect machining (assuming any significant machining would be required), in addition to subsequent mechanical damage in service. Today the fibre-resin composites are busy studying ways of repair from damage. Magnesium has 'repair' already well established.

The developments in future alloys from reinforced as well as other routes than the conventional melting route, are not really within the scope of this paper, but these will effectively eliminate the bogey of corrosion and provide even more competitive properties which would enable ever wider use of the new materials outside the applications described today.

My thanks for assistance in the preparation of this paper are due to the people referred to. The views expressed, unless specifically stated, are my own, based on the best information available to me.

REFERENCES

1   New frontiers in productivity with magnesium die castings, DWAIN MAGERS, Proceedings of the AFS/CMI special conference, June 19 1985

2   Magnesium supply and demand report, C W NELSON, 43rd World magnesium conference, International Magnesium Association

3   Materials for structures of the future, G B EVANS, Proceedings international conference, Materials in aerospace, The Royal Aeronautical Society, April 1986

4   Cost and weight considerations for different methods and suppliers, B SMYTHE, private communication

5   Thoughts upon the use of magnesium, G B EVANS, 28 Salon de l'aeronautique et de l'espace, Paris le Bourget 1969

6   Corrosion and design, G B EVANS, Royal Aeronautical Society, 1986

7   Reflections and projections (thoughts on the use of magnesium alloys in airframes), G B EVANS, 43rd World magnesium conference, International Magnesium Association

8   Developments in light metal forgings, E STEWART-JONES, Metal forming, November 1970

9   Evaluation of various magnesium finishing systems, A I ROBINSON, Proceeding of the AFS/CMI special conference 1985

Modern strategy for magnesium in automobiles: design, process and material aspects

N C SPARE

The author is in the Automotive Products Division of Kongberg Våpenfabrikk, Norway.

SYNOPSIS

For there to be a major shift of magnesium into structural usage the material must be accepted as a viable alternative to aluminium, zinc, steel and plastic by Automotive designers. The automotive engineer has a prerequisite to satisfy the function, reliability, quality and installed cost of components. The lightweight of magnesium is insignificant in itself and only when all aspects of the development and manufacturing route are fully integrated can the advantages of magnesium components be fully realised. Existing and future casting processes will provide improvements in properties but this must be backed up by further improvements in alloy performance and in the property/cost relationship.

INTRODUCTION

An observer looking at the end usage of magnesium could be excused for thinking that it is a metal struggling for an identity. It is a fact that about 80% of production goes either to other metals industries, for example in aluminium alloys, desulphurising of steel, nodularising of cast iron, for chemical reduction processes or in sacrifical applications. Whilst steel and aluminium have helped to shape our modern society, magnesium carries no such distinction being confined largely to playing a supporting role.

There are other metals for which this is the case, the so-called minor metals, but to apply this description to magnesium is to deny the very property which brought it to the fore 60 years ago, namely its lightweight.

It is also strange that magnesium based alloys have been used successfully in structural systems for 50 years and their quality and reliability is demonstrated every hour of every day in the aircraft industry.

A redistribution in the comsumption pattern of magnesium is long overdue and as the benefits in using the metal become better understood and if the industry responds to the market, the dependence of magnesium on other metals industries will assume less importance.

It is generally agreed that if the market for magnesium based alloys is to expand there must be acceptance by the automotive industry that this is a material which can make a positive contribution in vehicle designs of the future.

LIGHT WEIGHT AND ENERGY CONSIDERATIONS

What are the attractions of magnesium to automotive engineers? The primary reason is its lightweight; with a density of 1.8 gms/cm^3 it is the lightest structural metal. When alloyed it has useful mechanical properties and, dependent on the alloy system, sometimes exceptionally good properties. It has good castability which, when considered alongside its low specific heat and high thermal conductivity, makes it particularly attractive in certain casting processes such as high pressure die-casting.

The automotive industry is under pressure to reduce vehicle weight. The main driving force for this came in the wake of the energy crisis in the early 1970's. The US. government introduced legislation to ensure that the industry produced vehicles which would conform to fuel consumption limits (Corporate Average Fuel Economy). Fig. 1. The legislation is enforced by the Environmental Protection Agency (EPA) and manufacturers not meeting fuel economy targets can be penalised in the form of a tax (Gas guzzler tax).

Of course the automotive engineer has a number of solutions which can all contribute to improving fuel economy - downsizing, streamlining, increasing engine efficiency by improving combustion characteristics etc. All of these techniques have been used together with weight saving by the use of light weight materials such as magnesium [1]. Saving on vehicle mass is not the end of the story; whilst the industry is becoming more and more competitive, the customer is becoming more discerning. Automotive companies can only stay in the race if they are able to continuously improve quality, performance, comfort, fuel economy, smoothness and noise levels. Make no mistake; the automotive companies are battling it out for market share and there will be winners and losers.

MATERIAL CONSIDERATIONS IN PERSPECTIVE

The automotive industry is a giant and covers many disciplines. The main battle grounds are already drawn up. They are: Electronics, Manufacturing Techniques and Management Methods. The war of attrition will be fought in these areas and it is important to establish this perspective. Material and the choice of material is, in itself, a side-show, a Balkans campaign alongside Stalingrad or Normandy. The question is not whether it is possible to make the lightest, most streamlined, most comfortable car today, it is, can that car be made efficiently and be able to compete in world markets.

So far in this argument it has been established that the use of magnesium can contribute to weight reduction. By inference, if applied to components which reciprocate, oscillate or revolve it can also reduce the inertia forces in these dynamic components, thus contributing to better comfort and performance.

Having established the credentials of the material, that does not qualify it for battle. Only when the total process from raw material to final application is integrated in a cost effective manner to provide quality and reliability throughout the life time of the vehicle can magnesium become a candidate for substitution. The material must not be thought of in isolation but only as a part of a total design and manufacturing process which must be fully integrated if success is to be achieved.

Is there any evidence to suggest that magnesium components could meet these challenges in the late 80's and 90's ? The answer is yes and this will be examined to determine how strong the challenge is today, what lessons are to be learned from the past and how the industry will become stronger challengers in the future.

For the most important evidence we have to look to the USA. This is perhaps not surprising since the US automotive industry started from a very low corporate Average Fuel Economy in the 1970's. A significant number of new applications for magnesium parts have been established over the last 3-4 years, Figs. 2-7.

All of these parts, produced by high pressure die-casting, provide the carmaker with weight saving, manufacturing saving and functional improvements over other structural materials including plastics. The auto-manufacturer is able to benefit from the light weight metal and the high pressure die-casting process which permits the parts to be manufactured faster, more accurately, requiring less machining and with greater tool life than, for example, aluminium. But why should this happen now and not 15 years ago when the hot chamber die-casting process was first developed? The answer is not hard to find.

RESPONDING TO THE MESSAGE - IMPROVING CORROSION RESISTANCE

The automotive industry had a clear message - The corrosion resistance of magnesium castings was not good enough. In 1980 the message was finally received and the laboratories of Amax, the magnesium producer, set to work to see if the corrosion resistance of magnesium die-casting alloy AZ91 could be improved. The result was that within 4 years a magnesium die-casting alloy with corrosion resistance claimed to be as good as die cast aluminium 380 was available in the market at a price level no higher than the previous die-casting alloy [2].

As a result of good basic research which confirmed that by reducing impurity levels of iron, nickel and copper in the common die-casting alloy AZ91, corrosion resistance could be significantly improved, a field of automotive applications for magnesium high pressure die-castings was opened up. It should also be noted that the work has not stopped there and the other main producers Dow Chemical and Norsk Hydro are increasing their understanding of corrosion resistance of magnesium, extending the work into sand and permanent mould casting alloys and other commercial alloy systems. [3, 4, 5]

The work on improving corrosion resistance of magnesium alloys should be regarded as a very important milestone and has lead directly to a significant market break-through. It should be noted, without denigrating the work, that the effect of iron, nickel and copper impurities was identified and presented in a paper by Hanawalt, Nelson and Peloubet to the AIME in 1942. It hardly needs to be said that a 40 year reaction time is not good enough if a significant automotive market is to be established.

Perhaps there is a lesson to be learned from this. The period 1940-1960 was a kind of magnesium renaissance, enormous amounts of work on magnesium were carried out on both sides of the Atlantic and one wonders, how much other work, which could have significance today, is gathering dust in some dark and dusty archive? How much work was given up simply because the research tools, which today we accept as common-place, were simply not available?

The development of high purity alloys has served as a timely demonstration that the industry must be prepared to react to the messages coming from the market place. What other messages are being sent from the automotive companies.

DESIGN FOR MANUFACTURING, FUNCTION AND COST

Perhaps the most significant is contained in the keynote address given by G.F. Fred Bolling of the Ford Motor Company to the 1985 Scandinavian Aluminium Seminar.[6]
"The cost of material is only part of the real cost to the auto producer. The cost of the INSTALLED FUNCTIONAL FINISHED PART is the real cost.......... Poor quality adds to cost. We must never forget that good product design produces ease of manufacture, quality, and overall, the lowest cost".

In other words there are several inputs to each component in a vehicle. Material, design, process efficiency, quality, function and cost. Having accepted that the use of magnesium saves weight, the automotive engineer still remains to be convinced that the other key requirements can be satisfied.

The magnesium component supplier has therefore to present an overall design and manufacturing strategy. At the same time the barriers and misconceptions about magnesium must be overcome in order to establish a high level of confidence.

COMPUTER AIDED ENGINEERING

There is not today, in the automotive industry, a large pool of experience in designing magnesium components and in theory this presents a problem. However, with the advance of modern computer technology, at an ever reducing cost, techniques are available now which facilitate design optimisation and satisfy many of the automotive engineers prerequisites.

One such technique, gaining ground in the automotive sector is known as "Integrated Product Development" and is ideally suited to provide the design input which is today clearly lacking in the magnesium field.

At the onset it should be recognised that magnesium has properties which are its own. It is not a lighter version of

aluminium or zinc, it is unique, and this must be taken into consideration if components are to be designed to reflect those properties.

A typical product development programme begins with the customer when the function, load data and service conditions are established. By use of 2 and 3 dimensional modelling software a design data base can be established which provides the basis for both design optimisation and subsequent manufacture of models, prototypes, dies or patterns.

First, however, the component design can be analysed for deflections, stresses, interface conditions, resonance and temperature distribution by the use of Finite Element Analysis (FEM).

The method, using software developed to a high level of sophistication in the early 1960's as a result of the NASA space exploration programme, has a history of accuracy and reliability and is now becoming an established automotive design tool. Why is this technique of such importance when considering potential applications for magnesium ? The answer lies in the need for designs to take the particular properties of magnesium into account. Very often magnesium is dismissed without further consideration after the designer has familiarised himself with its Youngs' Modulus. When one has steel at 210.000 MPa, aluminium at 70.000 MPa compared to magnesium at 44.000 MPa, it is perhaps an understandable reaction to reject magnesium components on the grounds of having inadequate bending stiffness. This should not be allowed to happen because it is the combined effect of alloy property and component design which determines the bending stiffness. Since the bending stiffness is a function of the moment of inertia of a section and the moment of inertia increases with the cube of the section depth it can be seen that geometrical considerations are very significant in determining bending stiffness of a component.

Consideration of simple beams shows that magnesium is almost 19 times stiffer than steel on an equal weight basis. This has far more to do with the geometry of the beam section than any E differences, Fig.8.

In practice, for equal bending stiffness, we could expect to see far more ribbing on magnesium components than on components in aluminium or steel. Ribs are a very effective means of improving component bending stiffness without incurring an enormous weight penalty. The conventional wisdom is that magnesium components, having the same bending stiffness as components in aluminium, should show a weight saving of 20%-25%. By the same reasoning the figure will be a 65%-70% weight saving over steel.

In practical designing one of the benefits of the low E is that for the same load, magnesium will show a much better distribution of stress than steel or aluminium. This should enable the designer to further optimise the component weight since he can concentrate material more in areas of high stress and less in areas of low stress and even eliminate material where stresses are small. Figs. 9 and 10 give a demonstration of FEM stress analysis of a typical automotive component and shows the effect on stress distribution of different materials, in this case steel and magnesium, for the same loading, stiffness and boundary conditions.

Fig. 10 shows a magnesium bracket redesigned to give the same bending stiffness as the steel bracket shown in fig. 9. When loaded with the same force, the deflections will be the same. Examination of the stress levels shows a maximum of 279MPa in the steel and only 82MPa in the magnesium. If designing to yield with a safety factor of 2 it can be seen that a steel must be selected with a yield strength of 279x2=558MPa or a magnesium alloy with a yield strength of 82x2=164MPa, to give the same component performance.

This is a simple demonstration of how FEM can contribute to the process of designing automotive components. It will also tackle other areas where magnesium can give rise to difficulties, for example in components subject to thermal cycling when it interfaces with materials of different coefficient of expansion. At $26x10^{-6}/^0K$ compared to steel, $11x10^{-6}/^0K$ and aluminium $21x10^{-6}/^0K$, the high thermal expansion must be taken care of by establishing the necessary pre-stressing levels for room temperature assembly.

Whilst on the subject of thermal considerations, the technique has also been used to establish the temperature distribution in magnesium adiabatic engine components.[7]

Components can therefore be design optimised to an advanced stage before prototypes need to be produced and tested. It has not yet been suggested that the prototype and testing phase can be eliminated and probably it never will, however it is being claimed in the US that this technique can cut development time and cost by up to 90 %!

One of the attractions of the system is that having established the design database and optimised the geometry the information can be used, through a CAM system, to produce not only prototypes but also production tooling with any subsequent design changes being accommodated very rapidly and cost effectively.

Today, the tools are available in the form of computer aided engineering systems to enable the design of components with the most efficient and therefore cost effective use of material ..... or can it? Well, not exactly, for whilst the system is excellent it can only be as good as the data it is supplied with and here magnesium is not in particularly good shape. The fact is there is not enough material property data available. This is perhaps not true in the field of aerospace alloys but most of these alloys are precluded from the automotive area simply on the basis of cost. In the field of commercial alloys, well documented, reliable, process based data is decidedly lacking. At the same time, the automotive industry is not as good as it might be in being able to provide detailed loadings to which components are subjected in a vehicle lifetime. If minimum cost functional parts are to be produced it is essential to know, accurately, the functional requirements.

The lack of data for commercial magnesium alloys and processes has been mentioned and whilst it has been demonstrated how it is possible to overcome the fundamental problem of low E it has to be said that the property data that is available for the casting processes in use today does not provide us with a "leading edge" product. Compared to aluminium based alloys today's commercial magnesium alloys have inferior UTS, yield, fatigue, creep, and impact properties and suffer inherently from micro porosity.

Is there any hope for thinking that this situation can be improved so that magnesium components can be considered as applicable through the stress range established by the automotive industry? The answer is yes and lies in two areas - alloy improvement and casting process development.

ALLOY DEVELOPMENT

It has already been stated that in the area of aerospace applications magnesium alloys are available with exceptionally good properties. This is particularly true in elevated temperature environments and MEL have developed alloys with 250/250 tensile properties (250MPa at 250^0C) The history of MEL alloy development shows a continuous and impressive alloy performance improvement over the years well documented by S. B. Hirst [8], but the alloying elements eg. zirconium, silver, thorium, lanthanum, neodymium, etc. are expensive and processing costs high. However, serious efforts have been and are being made to successfully reduce the cost of these alloys. Whilst MEL have shown what can be achieved by way of alloy development their interest is in aerospace and speciality applications and will probably remain so. Is it not possible to expect a similar effort in the commercial alloy field or must the industry still work with an alloy, high purity of course, first developed more than half a century ago?

PROCESS DEVELOPMENT

The integration of product development is not complete unless all aspects of the manufacturing cycle are examined. Fig.11 examines a relationship between the casting process, the load requirements of the component, the mechnical properties and the degree of soundness produced by virtue of the process. There has been no attempt to quantify this relationship, it is simply to examine general tendencies. Within some obvious limitations this could be the picture for the same alloy system.

It is essential, when considering material properties, to relate these to manufacturing processes. Casting processes have been considered here since the bulk of magnesium automotive components will most likely be provided by this route. In the diagram, forgings have been included simply as a "marker" because of their characteristic soundness.

By this means it is possible to envisage a casting process strategy which matches the load requirements of components used throughout the automobile.

Some of the processes indicated here are already established for magnesium, the most notable being high pressure die-casting with many examples of successful application in covers, housings, grills, casings, etc. - generally parts subjected to quite low loading. For parts which are more highly stressed such as wheels, low pressure die-casting and high pressure die-casting with vacuum are established. Note that both of these processes permit the improvement of properties by heat treatment. So far as magnesium castings are concerned there is no other casting technique in use today providing regular production quantities of automotive components. Looking to the future perhaps there are lessons to be learned from developments which are taking place in the aluminium industry; these might give a pointer to where magnesium casting processes could lead in the future.

LOW PRESSURE SANDCASTING

Low pressure sand casting processes such as EDEM and Cosworth have demonstrated that when casting problems are identified and subjected to scientific research and when the results of those scientific findings are allied to innovative production engineering, the result is a casting process capable of producing components meeting the automotive industry's toughest requirements. The benefits of such a process in terms of its effect on improving casting soundness on magnesium alloys are now being fully evaluated and the indications are

that there are significant benefits to be gained by this route.

SQUEEZE-CASTING

In the area of components subject to high loads, both static and dynamic, perhaps also subjected to elevated temperatures, the combination of magnesium and the squeeze casting process holds out considerable promise. The manufacture of dynamic components in aluminium by squeeze casting is rapidly being established and will continue to gain ground. Squeeze-cast magnesium components are today under test in the automotive industry and there are good grounds for believing that highly stressed magnesium parts produced by squeeze casting will provide the automotive designer with an answer to reduce vehicle mass and inertia forces. It must not be forgotten that magnesium castings have been doing this for years in the aircraft industry - at a price. Today with squeeze casting the same can be done at a price and in volumes that are acceptable to the automotive industry.

In terms of process strategy the most important factor in the final analysis is cost. An attempt has been made to superimpose a broad, relative process cost relationship to the diagram, but the cost picture is very complex and depends on a number of factors such as volumes, complexity of design, size, opportunity for multi-impression tooling or moulds etc.

RAW MATERIAL - THE COST OF PROPERTIES

The means are available today for material process and design optimisation to ensure that magnesium components can meet the customers' requirements for quality, functional effectiveness and cost. In the total cost picture, however, the raw material price has not been discussed. Whilst, as Bolling stated, this is only part of the real cost to the automaker; in the total process to installation and function it is probably the largest single cost factor. There is much talk about the aluminium/ magnesium price ratio but in an industry as competitive as the automotive industry its use is devalued since it has been shown that it is a combination of material property and design optimisation which determines the weight of a component. Direct substitution of, for example, diecast magnesium for diecast aluminium does not ensure equal component performance. This does give rise to confusion and has been responsible for the magnesium industry scoring own-goals when it simply relates properties to density without accounting for design optimisation factors.

Perhaps the best way to look at this is that the automotive industry does not really care if a component is magnesium or aluminium or plastic. It is in the business of buying properties and function. The magnesium industry would do well to remember this. The cost of properties in the commercial alloys such as AZ91 are not competitive. This alloy is not a direct substitution for aluminium alloy such as LM24 on a volume basis. For equal component performance there is not a 30% weight saving. Dependent on which property the engineer is designing for the weight saving can be no more than 25% or as little as 14% if designing for fatigue. The development of processes to improve casting properties is going on apace, it will not alone make magnesium castings competitive enough for the automotive industry. There are two simple choices for the magnesium producers if a thriving automotive market is to be established - Get the price down or the properties up ! Preferably both !

NEW MAGNESIUM PRODUCTION CAPACITY

The grounds for optimism on alloy improvement have been mentioned; on the supply side there are also some optimistic developments. Firstly, there have been considerable improvements in electrolytic extraction technology which has lead to greater efficiency and better cost effectiveness of the process. Norsk Hydro in Norway have been in the forefront of this development. Secondly, a new route for the production of magnesium chloride has been developed by Mineral Process Licensing Corporation (MPLC) offering the prospect of a considerable reduction in process energy cost. Thirdly, the construction of a total of 7 new magnesium extraction plants has been announced in the past 12 months with a combined capacity of 250,000 tonnes. When we look at existing free world capacity of 225,000 tonnes, the students of supply/demand economics can work out what that means.

It would be naive to expect that in three years all of this capacity will come on stream. Nevertheless it is a demonstration of confidence in the future of magnesium and a clear signal to the automotive industry that the raw material will be available at a price it can afford.

CONCLUSION

It has been shown in this paper that new technology and the pace of change is ever increasing. If the new technological resources can be harnessed by the magnesium industry and targeted towards the automotive manufacturers so that they are no longer told that magnesium is the lightest structural metal and expected to buy it, the metal has an opportunity, like never before, of playing an established role in the modern road vehicle.

ACKNOWLEDGEMENTS

The author wishes to thank the management of the Automotive Products Division of Kongsberg Våpenfabrikk for their kind permission to give this paper and to his colleagues for their invaluable contribution.

REFERENCES

1. George B. Kenney: "Energy Features of Magnesium Production and Use in Automobiles". SAE Paper 820152. February 1982.

2. H.J. Dundas: "AMAX Speciality Metals Corporation". Report ISJ-1324.

3. James E. Hillis: "The Effect of Heavy Metal Contamination on Magnesium Corrosion Performance". SAE Paper 830523. February 1983.

4. K.N. Reichek, K.J. Clark and J.E. Hillis: "Controlling the Salt Water Corrosion Performance of Magnesium AZ91 Alloy". SAE Paper. February 1985.

5. J.E. Hillis and K.N. Reichek: "High Purity Magnesium AM60 Alloy: The Critical Contamination Limits and the Salt Water Corrosion Performance". SAE Paper, February 1986.

6. G.F. Bolling: "Some Remarks About Aluminium's Share of Automotive Materials". Skandinaviske Aluminiumdager '85 - Copenhagen, 2nd - 3rd September 1985.

7. P. Glance: "Computer Aided Engineering Analysis of Magnesium Adiabatic Engine Components". International Magnesium Association. Annual Meeting Paper 1985.

8. S.B. Hirst: "50 Years of Magnesium - The MEL Story". International Magnesium Association. Annual Meeting Paper 1984.

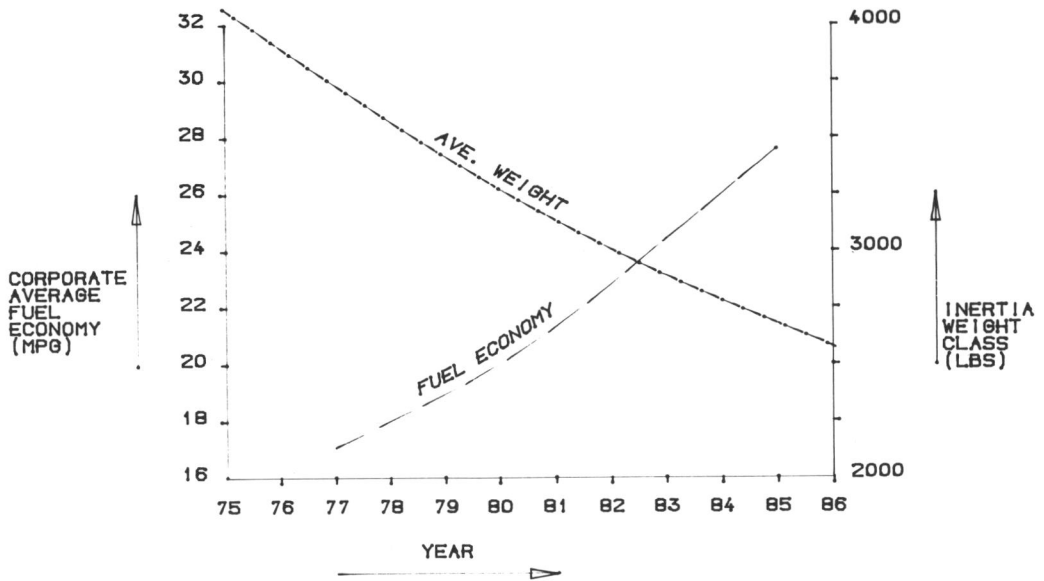

1 Vehicle weight/fuel consumption
historical development, USA

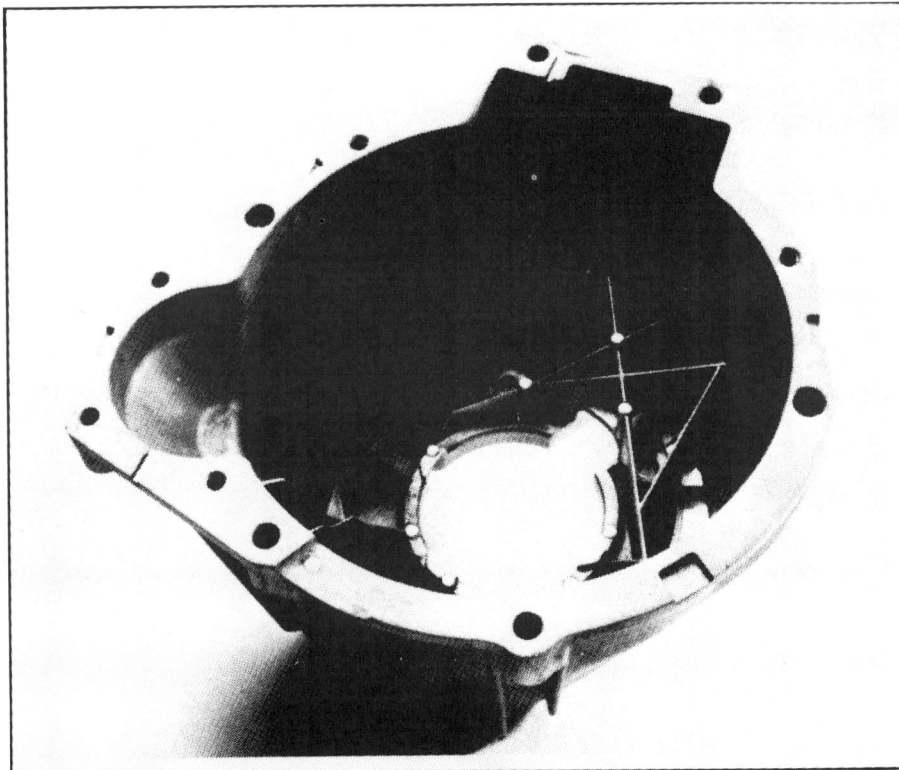

2 Clutch housing, Ford Light Trucks

3   1984 Chevrolet Corvette air cleaner cover

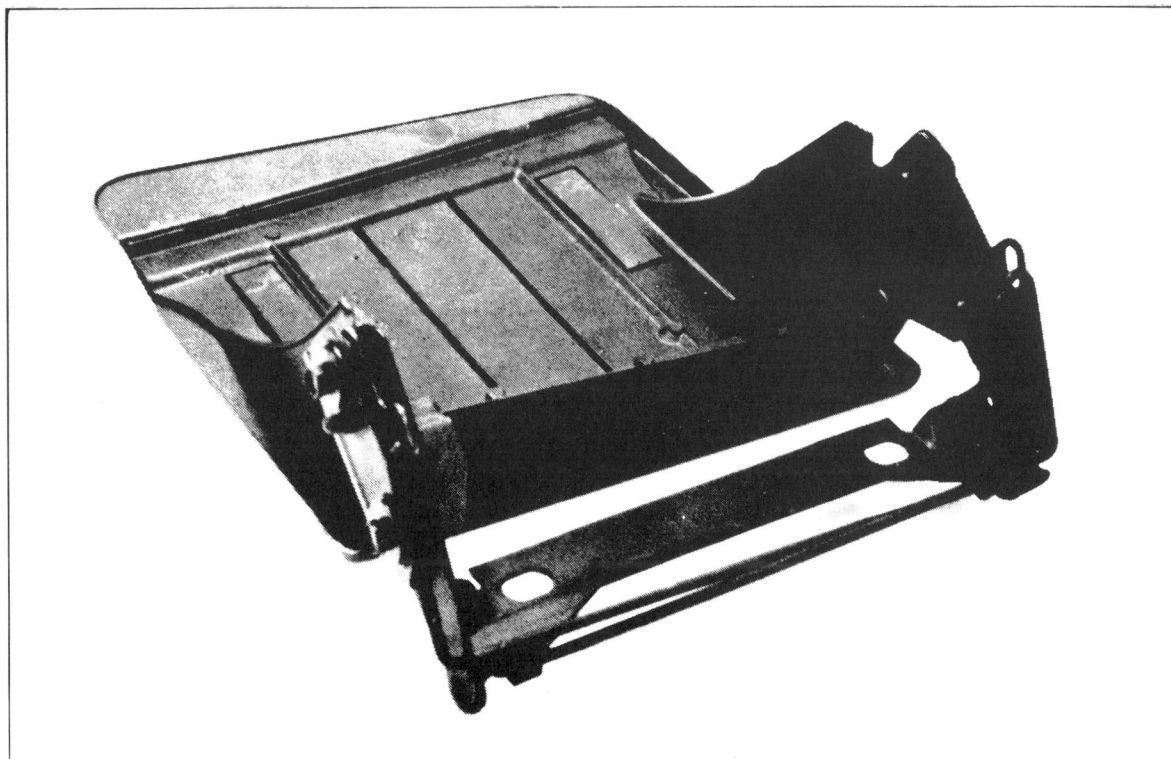

4   Pontiac Fiero headlamp door

5 Rocker arm cover for Chevrolet Corvette

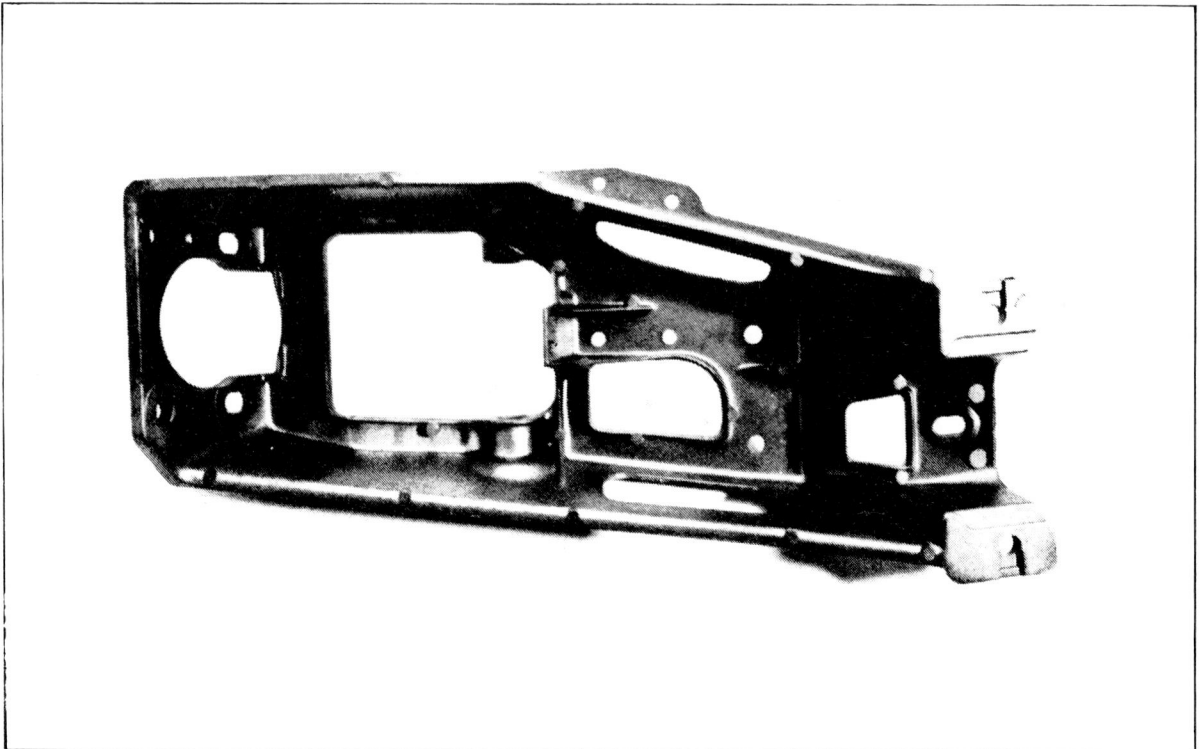

6 Clutch and brake pedal support bracket
   for Ford Trucks

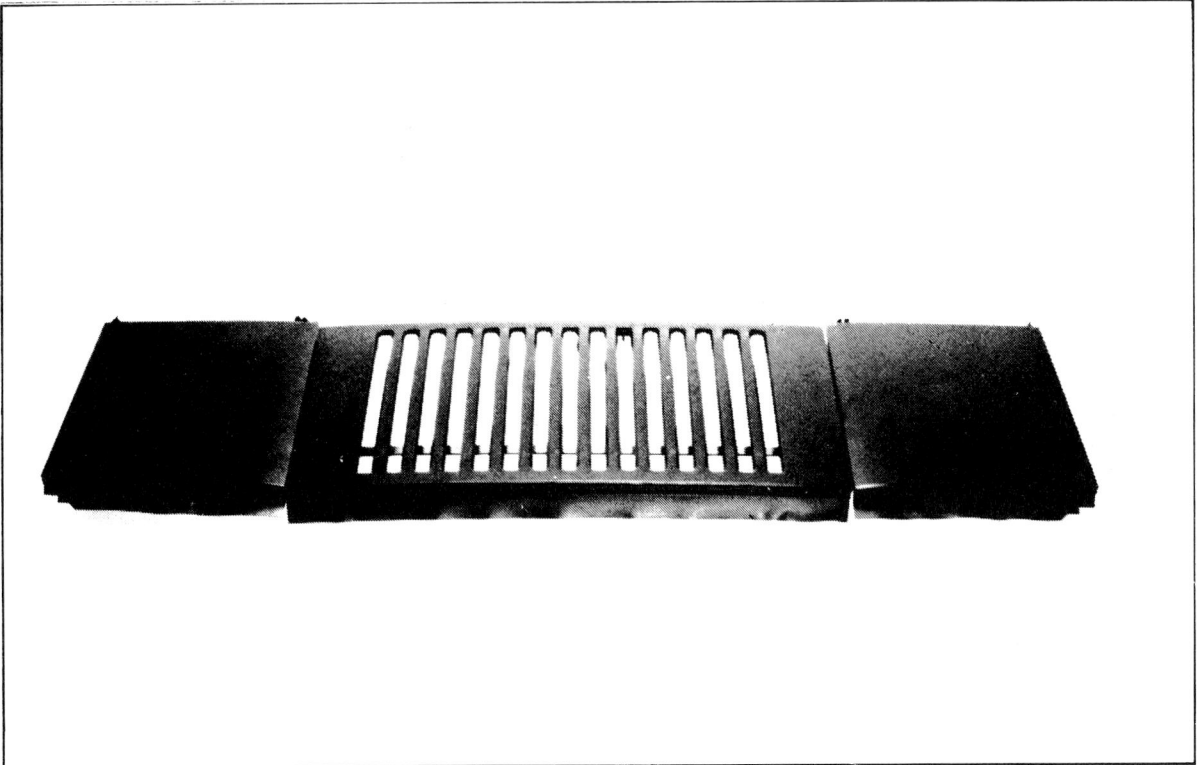

7 Pontiac Fiero three-piece engine
   compartment grill

STIFFNESS

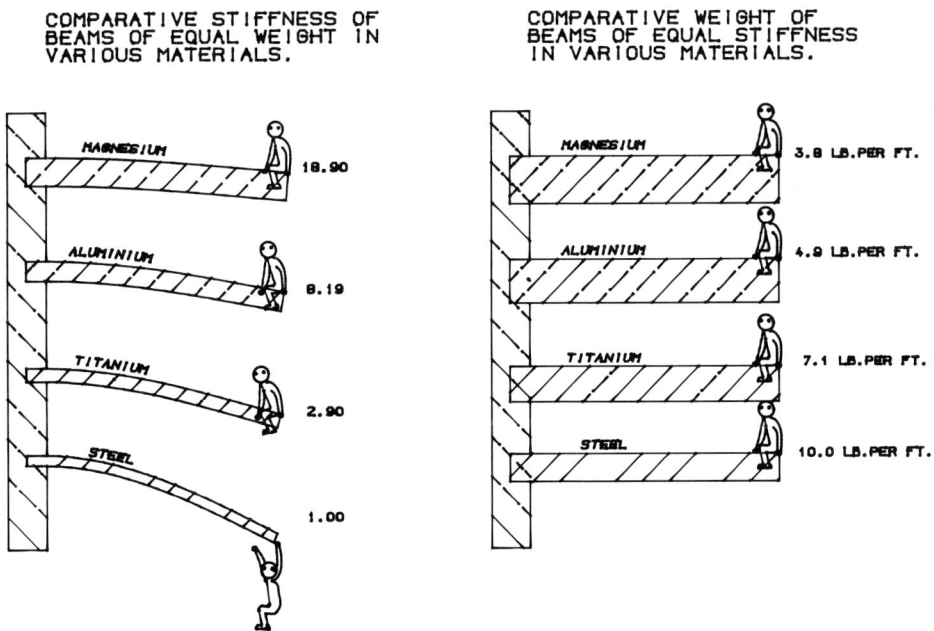

COMPARATIVE STIFFNESS OF
BEAMS OF EQUAL WEIGHT IN
VARIOUS MATERIALS.

COMPARATIVE WEIGHT OF
BEAMS OF EQUAL STIFFNESS
IN VARIOUS MATERIALS.

MAGNESIUM   18.90
ALUMINIUM   8.19
TITANIUM   2.90
STEEL   1.00

MAGNESIUM   3.8 LB. PER FT.
ALUMINIUM   4.9 LB. PER FT.
TITANIUM   7.1 LB. PER FT.
STEEL   10.0 LB. PER FT.

Figure 8

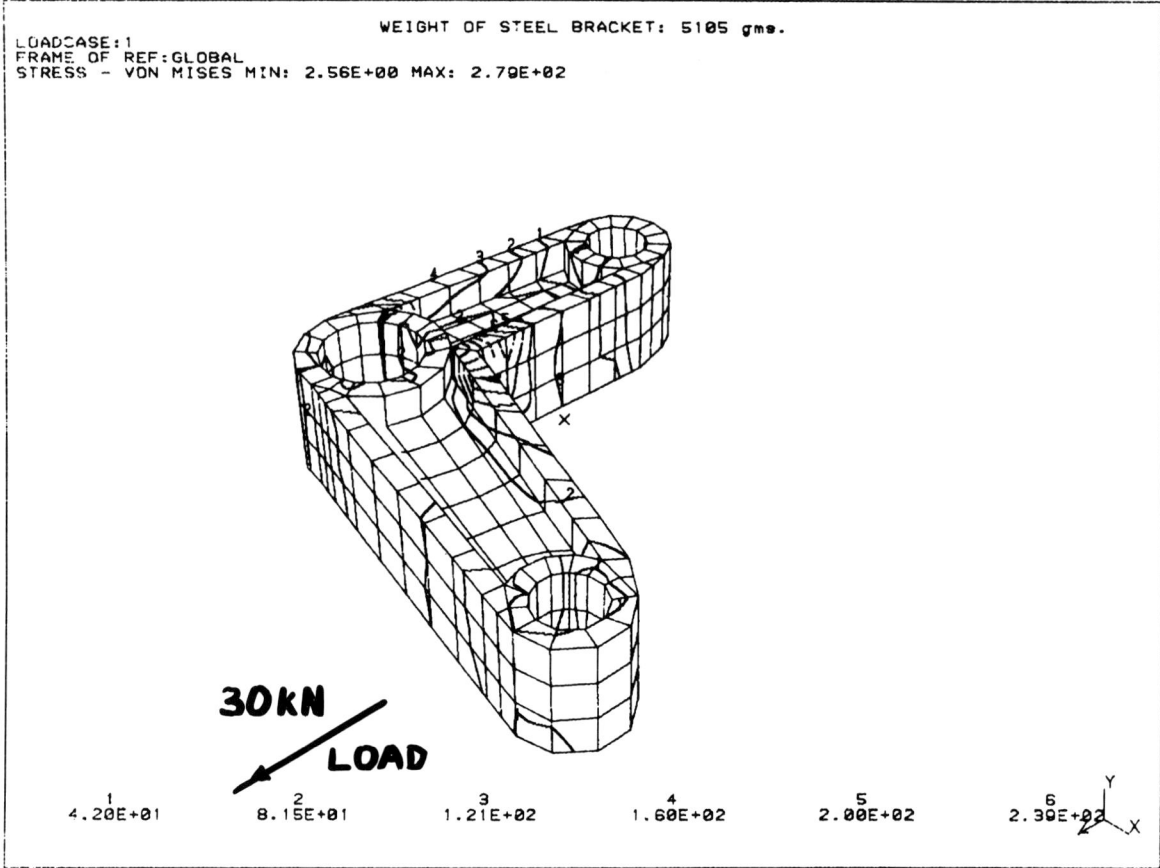

Figure 9

WEIGHT OF MAGNESIUM BRACKET: 1554 gms.

LOADCASE:1
FRAME OF REF:GLOBAL
STRESS - VON MISES MIN: 3.50E+00 MAX: 8.22E+01

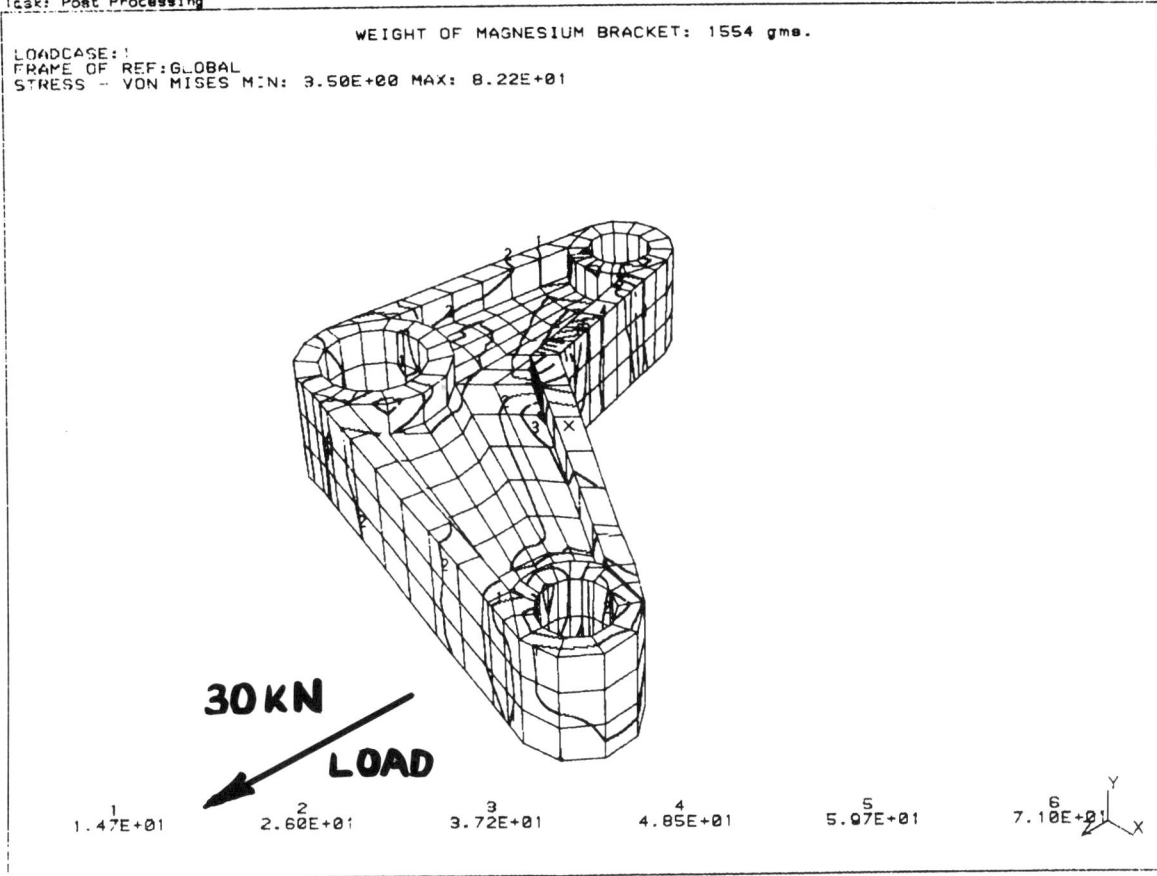

30KN

LOAD

1	2	3	4	5	6
1.47E+01	2.60E+01	3.72E+01	4.85E+01	5.97E+01	7.10E+01

Figure 10

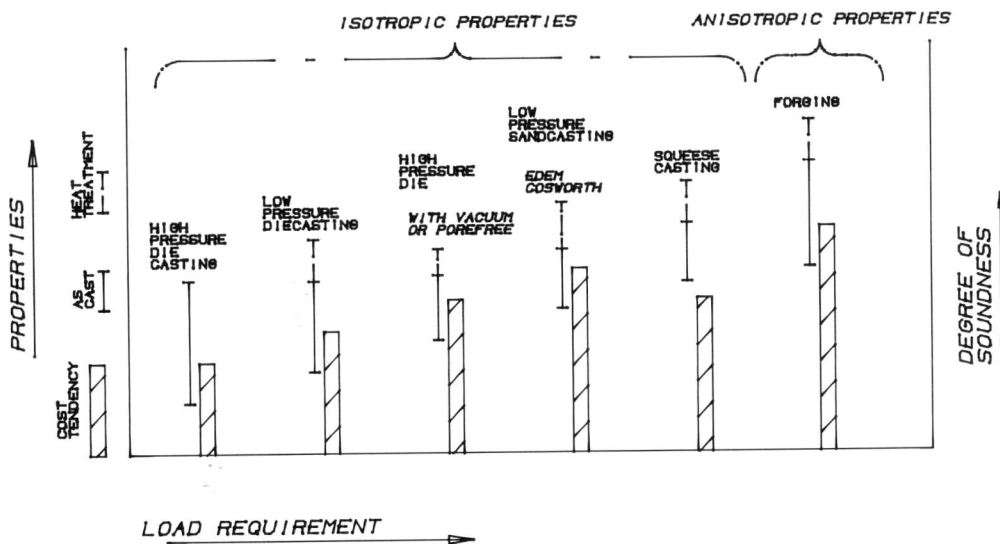

11 Process/property relationships

Magnesium in British nuclear industry

W G SHAKESPEARE and K G SUMNER

Mr Shakespeare is with the Fuel Element Design Office of British Nuclear Fuels plc, Preston, England. Mr Sumner was until retirement in a similar position with BNFL and is now a part-time consultant to Fairey Nuclear Limited, England.

SYNOPSIS

The Mark I Gas-Cooled Thermal Reactor System (Uranium/Magnox) has formed the basis of the Nuclear Power Industry within the United Kingdom. Though now being replaced by the Advanced Gas-Cooled Reactor AGR, the system, which uses metallic uranium encased in a magnesium alloy envelope, has operated successfully for the past 30 years. The paper reviews the use of magnesium in the nuclear industry both as raspings for the reduction of uranium tetrafluoride to produce metallic uranium and in the series of alloys (now widely known as Magnox alloys) used in the manufacture of fuel cans and other components needed for the production of fuel elements.

INTRODUCTION

Towards the end of the second World War, the British Government initiated the building of two atomic piles at Windscale for purely military purposes. These used metallic uranium rods contained in an aluminium envelope and gas cooled. It soon became evident that these piles were capable of improvement and could be made to produce power as a by-product to the prime military application and, in the late 1940s, the Government sanctioned the building of four nuclear reactors-christened PIPPA – at Calder Hall adjacent to the Windscale site. Subsequently it was decided to build a further four similar reactors at Chapelcross in southern Scotland. The reactors were again based on metallic uranium rods and gas cooled but the envelope material was changed to magnesium. In the mid 1950s, a decision was taken to build three commercial nuclear power stations, based on the PIPPA design but with considerably enhanced electrical power output. These three power stations, each with two reactors, were commissioned over the period 1962-64 and, whilst they were being built, a further six twin-reactor stations were ordered in Britain. Orders for two single-reactor stations abroad, one in Italy and one in Japan, were also received. All stations were operating by 1971. The UK Power Stations, together with their electrical outputs and dates of commissioning are listed in Table 1.

PRODUCTION OF URANIUM

In the production of uranium, the ore is converted into uranyl nitrate which is purified and converted into uranium tetrafluoride. The uranium tetrafluoride is intimately mixed with magnesium, in the required proportions and the mixture pressed into compacts. The compacts are stacked in reaction vessels, heated and the uranium tetrafluoride reduced by the magnesium to yield pure metallic uranium.

The magnesium is used in the form of raspings and the shape and packing density of these is of significant importance in maintaining the efficiency of the reduction process. Similarly, since more than 80% of the impurities present in the magnesium are transferred to the uranium, the composition of the raspings (Table II) must be carefully controlled to tight limits if the stringent composition requirements for the uranium are to be achieved.

Since, to improve fuel rod stability in the reactor, a small addition of aluminium is made to the uranium, more recent development work has enabled the use of a small proportion of high quality turnings, produced during on site machining of Magnox AL80 fuel cans, to be used along with the high purity raspings in the production of compacts.

CAN AND COMPONENTS

There are three major requirements for fuel element components:

1. They must be of high quality and not prone to failure. This is particularly so for those components which form the fuel envelope, and for such components it is usual to call for searching non-destructive testing on both material and finished product.
2. They should not be difficult to manufacture to a consistently high standard that is amenable to quality control inspection techniques.
3. Since many hundreds of thousands of components are required during the life-time of a reactor, they should be as cheap as possible without sacrificing the over-riding requirements for soundness.

TABLE I

UK URANIUM/MAGNOX POWER STATIONS

STATION	No REACTORS	STATION OUTPUT	DATE COMMISSIONED
CALDER HALL	4	200 MwE	1956
CHAPELCROSS	4	200 MwE	1959
BERKELEY	2	276 MwE	1962
BRADWELL	2	300 MwE	1962
HUNTERSTON 'A'	2	320 MwE	1964
HINKLEY PT 'A'	2	500 MwE	1965
TRAWSFYNYDD	2	500 MwE	1965
DUNGENESS 'A'	2	550 MwE	1965
SIZEWELL	2	580 MwE	1966
OLDBURY	2	600 MwE	1968
WYLFA	2	1180 MwE	1971

TABLE II

COMPOSITION OF MAGNESIUM RASPINGS

ALUMINIUM	250 ppm	LEAD	50 ppm
BORON	0.3 ppm	LITHIUM	2 ppm
CADMIUM	0.3 ppm	MANGANESE	60 ppm
CHLORINE	100 ppm	NICKEL	20 ppm
CHROMIUM	20 ppm	NITROGEN	200 ppm
COPPER	20 ppm	SILICON	75 ppm
IRON	200 ppm	RARE EARTHS	*

*180 (ppm Gd) + 26 (ppm Sm) + 19 (ppm Eu) +
+ 4 (ppm Dy) + 0.3 (ppm ORE) not greater
than 50

TABLE III

COMPOSITION OF MAGNOX ALLOYS

ALLOY ELEMENT	AL80	MN80	MN150	ZR55
ALUMINIUM	0.7–0.9%	500	500	200
BERYLLIUM	0.002–0.030%	–	–	–
CALCIUM	80	80	80	80
COPPER	100	100	100	100
IRON	60	300	300	60
MANGANESE	150	0.70–0.90%	1.30–1.70%	150
NICKEL	50	50	50	50
SILICON	100	200	200	100
THORIUM	1	4	4	4
ZINC	100	300	300	150
ZIRCONIUM	250	250	250	0.45–0.65%

LIMITS ARE IN ppm UNLESS OTHERWISE STATED
SINGLE VALUES INDICATE MAXIMA

a. Materials

There are six major factors which influence the choice of materials for use in nuclear reactors. These are:

1. compatibility with the fuel;
2. compatibility with the reactor coolant;
3. low rate of corrosion in cooling ponds;
4. low neutron absorption cross section;
5. adequate combination of strength and ductility; and
6. behaviour under irradiation

When the Uranium/Magnox reactors were first planned, the commercially available magnesium alloys were assessed against the above criteria and all but one (AM503S) were found to be unacceptable for various reasons, and even AM503S did not fully comply with all these requirements. A joint development programme, between UKAEA Harwell and Magnesium Elektron Limited, was undertaken to produce an alloy which had a high resistance to oxidation in moist air and carbon dioxide and an ignition temperature which approached its melting point. Initially an alloy, Magnox E (the term was derived from Magnesium No Oxidation), which was a Mg-1% Al 0.05% Ca 0.01% Be alloy was suggested. Difficulties in obtaining gas tight sealing welds and failure to pressurise the can down onto the uranium bar at the desired temperature/time/pressure conditions were encountered and were overcome by reducing the aluminium content and eliminating calcium. Eventually a whole series of Magnox alloys was developed. These include Magnox AL80, an alloy containing 0.8% Al and 0.01% Be, for the manufacture of fuel cans; Magnox MN80 and MN150, containing respectively 0.8% and 1.5% manganese, used for end caps and end fittings where high creep strength is required; and Magnox ZR55, a magnesium alloy with 0.55% zirconium, used for structural components on the fuel element where good creep strength allied to low neutron absorption cross section is necessary. Full chemical compositions of these alloys are detailed in Table III. In the development of these alloys, the research staffs of the AEA and the reactor building consortia undertook considerable programmes of work, much of which has been reported in various places over the years. In addition considerable assistance was given by both Birmetals Limited and Magnesium Elektron Limited and all the Magnox alloys, for initial and replacement reactor charges, were supplied by one or other of these two companies. Since the demise of Birmetals, Magnesium Elektron Limited has become the sole supplier.

Magnox AL80 was chosen for cladding because at reactor temperatures (typically 200–500°C ie up to 85% of the absolute (°K) melting point of the material) it has sufficient ductility to accommodate the strain imposed by dimensional changes in the fuel rods, combined with adequate creep strength to prevent distortion of the fins by the turbulent gas flow necessary for maximum heat transfer.

The raw material, used for the production of Magnox AL80, has been, traditionally, thermally reduced magnesium to a carefully controlled purity. Trials using electrolytically reduced metal resulted in a higher than normal in-reactor failure rate. This was considered to be associated with a higher than usual iron content and modified melting techniques were introduced to give analyses, for the major impurities, similar to those for metal from thermally reduced magnesium. However there is still some doubt that, under certain reactor conditions, such material behaves exactly like its thermally reduced counterpart. No such problems exist with the production of other Magnox alloys.

On independently supported fuel elements, Magnox AL80 is also used for the manufacture of end caps, but on stacked designs, where each fuel element is supported by those below it in the channel, a more creep resistant material is required to limit compression of the end cap base and consequent deformation of the can. The binary magnesium – manganese alloys were chosen for this application and Magnox MN80 was used for the earlier reactors, but Magnox MN150 has been introduced for the later reactors where the operating conditions are more onerous. The same alloys are used for the manufacture of end fittings. To attain the required level of creep strength, the component or material is heat treated to coarsen the grain size to a minimum of 0.5 millimetre. Although the neutron absorption cross sections of these two alloys are respectively 70% and 130% greater than that of Magnox AL80, the weight of material involved is relatively small and the penalty on neutron economy has to be accepted.

For splitters and braces, which locate the fuel elements centrally in the reactor channels and are designed to restrict the bowing of the fuel elements, Magnox ZR55 was chosen. Normally this alloy exhibits a very small grain size of about 50 microns, but this can be increased to 250–500 microns, with consequent improvement in creep strength, by heat treating for up to 72 hours at 600°C in carbon dioxide. The creep strength can be still further enhanced by producing a fine precipitate of zirconium hydride in the material. To achieve this a small amount of water vapour is introduced into the heat treatment atmosphere.

Magnox ZR55 was also considered as a canning material, but it was found that, at the higher temperature positions, plutonium was able to diffuse through the can wall and into the reactor atmosphere, thereby seriously affecting the sensitivity of the burst cartridge detector gear, and its use for this purpose was abandoned. Arising from this it was established that although plutonium is injected into the bore surface of Magnox AL80 cans, it does not penetrate more than a few microns into the can as it combines with the aluminium in Magnox AL80 to form a compound which becomes locked in the magnesium lattice. It has been established that such plutonium diffusion through Magnox ZR55 cans does not occur in the lower temperature positions of reactors. The possibility of using a Magnox ZR55 can, which will have both a higher creep strength and a more stable fine grain structure than its Magnox AL80 counterpart, in the lower temperature positions is still being considered.

b. Manufacture

The cans used for the original Pippa fuel elements in the reactors at Calder Hall and Chapelcross were machined from an extruded solid bar of Magnox AL80. The fins were

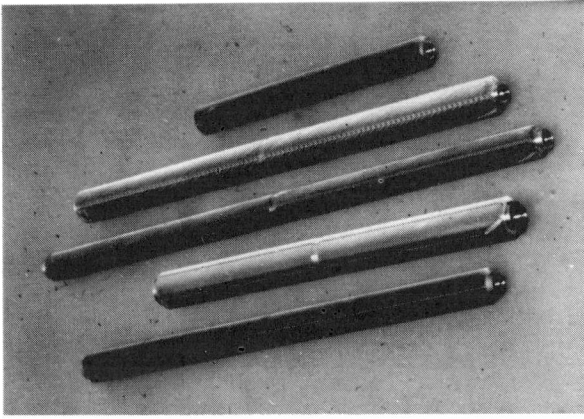

Figure 1 – Selection of Helical Cans used in the earlier Reactors.

Figure 2 – Herringbone Can and Cruciform extrusion from which it is machined.

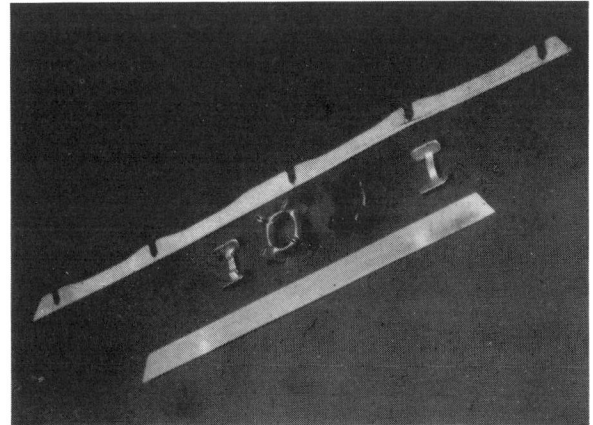

Figure 3 – Selection of End Caps.

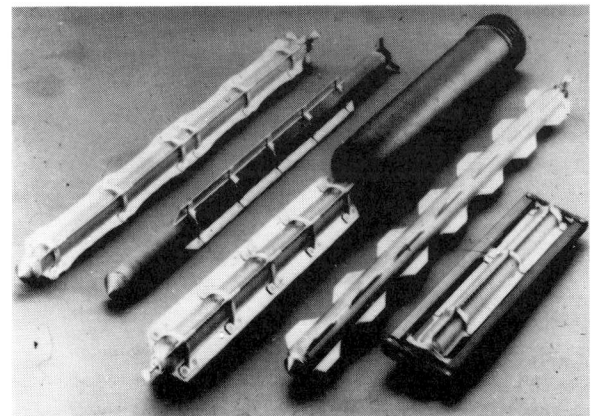

Figure 4 – End Fittings for various types of fuel elements.

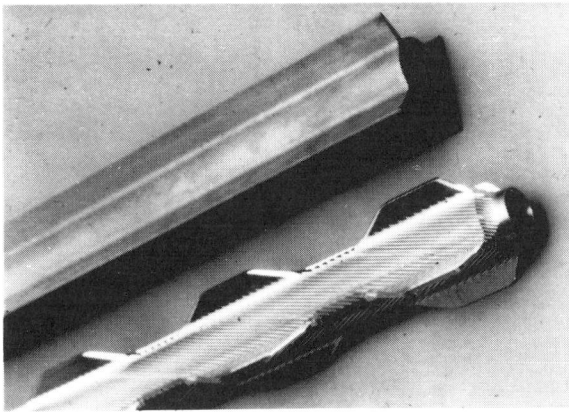

Figure 5 – Splitter Blades and Braces used for constructing the anti-bowing device on fuel elements.

Figure 6 – Selection of Fuel Elements.

virtually circumferential, being single start on a 3 millimetre pitch helix. The overall diameter of the can was 54 millimetres and it accommodated a fuel rod 29.2 millimetres diameter and just over a metre in length. The wall thickness of the can was 1.5 millimetres. The bar from which the cans were machined was subjected to 100% ultrasonic inspection and, in addition, samples were taken from the extrusion billets and at intervals along the extruded bar for metallographic examination for harmful defects such as flux or other inclusions, cracks etc.

Later designs of Pippa can had true circumferential finning. These were manufactured from extruded, thick-walled Magnox AL80 tubing by a roll finning process in which the heated tube passed progressively through a series of finned rolls, the fins on each successive roll being deeper, so that eventually a fin of the desired form was thrown up. The overall dimensions of this can were the same as the earlier design and the extruded tube was subjected to the same stringent inspection.

For the manufacture of polyzonal cans (Figure 1), which have multi-start helical finning and were developed for use in the commercial Uranium/Magnox power stations, extrusion slugs are machined from extruded Magnox AL80 bar. These are re-extruded into longitudinally finned tubes which are subsequently heated and twisted to the required helix. Designs in current use have 40-72 fins, the helices range from 0.3 to 0.45 metre and the wall thickness from 1.5 to 2 millimetres. The lengths of the fuel rods range from 0.48 to 1.07 metres and the diameters of most are between 28 and 29 millimetres, but there is one design at 41 millimetres diameter. Again, the extruded bar is subjected to rigorous ultrasonic testing and micro-examination.

The latest development in the Magnox can designs is the herringbone (Figure 2) in which the fins are arranged in a chevron pattern with four integral lugs. These cans are machined from extruded, thick-walled hollow cruciform sections in Magnox AL80, the external dimensions of the extrusion being held to those required on the finished can. The extruded section is subjected to the usual ultrasonic testing and micro-examination. Cans of this configuration are now used in the eight reactors at Calder Hall and Chapelcross and in eight reactors at commercial power stations.

In the early life of Pippa reactors, fuel was seldom irradiated for longer than one year and there were very few instances of can failure. However, in preparation for the civil power programme, longer irradiations were undertaken and it soon became apparent that the rate of failure in the lower (colder) element positions was increasing. Examination of these failures showed that strain induced in the can, by dimensional changes in the fuel, coupled with the effects of irradiation, had resulted in the formation of cavities at the grain boundaries, and, with the relatively coarse grain size of the cans (around 0.5 millimetre), these cavities were linking up to form leak paths through the can wall. To overcome this, manufacturing routes were developed for polyzonal cans in which the grain size of the can on the finished fuel elements was restricted to a maximum of 0.25 millimetre, and irradiation trials showed that such cans were not subject to cavitation failure. However, when such fine-grained cans were irradiated in the upper (hotter) channel positions, marked grain growth occurred and instances were observed where single grains occupied the whole can wall thickness and extended 50 millimetres or more along the can. Such structures were considered to be undesirable and to prevent grain growth to this extent in the hotter positions (ie above 350°C), an initial grain size of around 0.5-0.25 millimetre is required. Manufacturing routes have been established for cans which, on the finished fuel element, comply with this specification.

With most designs of polyzonal can, it has been possible to achieve the desired finished fuel element can grain structures using identical can manufacturing routes followed by an appropriate fuel element pressurising route. For the low temperature positions (fine-grained), the cans are hydraulically pressurised at 250°C onto the fuel rod followed by a half hour anneal at 390°C. For the high temperature positions, the cans are annealed for 30 minutes at 320°C prior to use and the sealed cans gas pressurised at 510°C. With some designs, significantly different can extrusion and/or twisting conditions have been necessary to produce low and high temperature cans. For herringbone cans, two completely different manufacturing routes are required to produce cruciform extrusions which give the desired grain structure of cans after pressurising. Pressurising, in addition to achieving a reasonably stable grain structure of the required size, pushes the can down onto the fuel rod and into grooves machined in it. This overcomes the problem of the differing expansion characteristics of magnesium and uranium, thus allowing both can and fuel rod to move together during the various thermal cycles encountered during reactor operation.

The end caps used on Uranium/Magnox fuel elements are generally of a simple cup shape, but in some designs this is complicated by the presence of a centre stud used for the attachment of end fittings or supports (Figure 3). One of the main objects in the manufacture of an end cap is to ensure that, as far as possible, there is not a potential leak path through the wall or base. Since non-metallic inclusions may occur even in the most carefully controlled melts of magnesium alloys and might, therefore, be found in an end cap, insistence is placed on the attainment of near perfect U-shape grain flow in the component.

End caps are machined from forgings made from extruded bar in Magnox AL80, MN80 or MN150 which has been subjected to the same stringent inspection procedures as that used for the manufacture of cans. The magnesium-manganese alloy forgings are heat treated to give a mean grain size of 0.5-1.5 millimetres. Samples from each heat treatment batch of forgings are checked for compliance with these grain size requirements and also for the adequacy of grain flow, and every end cap is ultrasonically tested to detect any inclusions which might have arisen during the forging operations.

Stacked fuel elements are fitted with top and bottom end fittings (Figure 4). The top fitting usually has a conical cup shape into which the cone bottom end fitting of the next element up the channel locates. These fittings carry the whole weight of the elements and consequently require good creep strength which is achieved by use of heat-treated Magnox MN80 or MN150 alloys.

Top end fittings are of three forms. A simple cup, a cup with three fixed arms which locate the element centrally in the channel, and a spring-arm variety in which one of the fixed arms is replaced by a spring-loaded hinged arm designed to damp out vibration effects. The simple cup shape is machined from heat-treated bar; the three arm type is machined from heat-treated die forgings; and the spring-arm type is machined and assembled from a combination of die castings and heat-treated forgings. The required grain size in the castings is attained by control of the casting process and cooling rate.

Bottom end fittings are machined from a heat-treated extruded bar. The cone face of the bottom end fittings is sheathed in stainless steel to prevent undue wear, and the mating surfaces of the cups and cones are coated with colloidal graphite to prevent them sticking together in the reactor.

Independently supported fuel elements use a variety of end fittings in stainless steel and zirconium-base alloy. Only the top end guide of the Berkeley fuel element is made in magnesium alloy (Magnox MN80).

As mentioned earlier, splitters are attached to the fuel elements to locate them centrally in the channel and to prevent bowing arising from creep or irradiation effects. To achieve the necessary strength, these are manufactured in Magnox ZR55. They are extruded as plain strip, typically 2 millimetres thick, and this is heat-treated to produce the required creep strength. The plain strips are finally machined to the required splitter shape (Figure 5).

Splitters are attached to the fuel elements by braces which are also manufactured in Magnox ZR55. These may be simple forged W-braces or more complex ring-braces which are machined from either hollow cruciform extrusions or die forgings (Figure 5). All braces are subjected to the special heat treatment devised for this alloy and referred to earlier.

In the assembly of fuel elements, end caps are fusion welded to the cans, splitters and braces are resistance spot welded into position and end fittings are attached mechanically. A selection of fuel elements is shown in Figure 6.

GENERAL COMMENTS

Up to the end of October of this year, BNFL had manufactured over 3.75 million Uranium/Magnox fuel elements and of those which have been irradiated less than 0.001% have failed in service. This is the result of the care and attention paid to quality control procedures and the rigorous inspection standards applied during the manufacture of components and assembly of fuel elements.

Furthermore, when the commercial Uranium/Magnox power stations were designed, the anticipated burn-up of the fuel was about 3000 MWd/t. Now we are achieving channel average irradiations between 5000 and 6000 MWd/t with peak element irradiations well in excess of 7000 MWd/t.

The Uranium/Magnox power stations have been undoubtedly successful. For many years all 9 stations figured in the 'top twenty' of the World's nuclear power stations in terms of generated electrical output. Generating costs have been approximately 20% less than those for coal-fired stations. However their role has been to meet base load requirements and, in the United Kingdom, the system has been surplanted by the more versatile and efficient Advanced Gas-cooled Reactor system (AGR) which uses uranium oxide fuel in a stainless steel envelope. It is expected that the turn of the century will see the end of this unique use of magnesium.

ACKNOWLEDGEMENTS

The authors wish to express their thanks to the Directors of BNFL for permission to publish this Paper and to several of their colleagues for assistance in its preparation.

Magnox oxidation in carbon dioxide

H M FLOWER and A J MORRIS

Dr Flower is in the Department of Metallurgy and Materials Science, Imperial College, London.
Dr Morris is with the Olin Corporation, Metals Research Laboratories, Newhaven, USA.

SYNOPSIS

The mechanism of oxidation of magnesium alloyed with aluminium (0.8 wt%) and beryllium (0-160 ppm) has been studied in moist carbon dioxide at 550 C. In the absence of beryllium oxidation is associated with breakdown of the initially formed compact oxide and breakaway involving vapour transport and oxidation of magnesium to produce a porous oxide layer. Gas transport through this layer results in a high CO/CO_2 ratio at the metal/oxide interface and the Boudouard reaction takes place forming carbon. This diffuses into the metal to form a mixed carbide, approximating to $Al_4Mg_2C_3$. The local depletion in aluminium causes breakdown of the planar oxidation front and the formation of deep spike shaped, oxide filled, intrusions into the metal in grains where the basal plane is inclined to the surface. Beryllium additions progressively reduce the oxidation rate and increase the resistance to breakaway. If beryllium is present in sufficient quantity beryllia forms at the metal/oxide interface and continues to grow in preference to MgO. This limits continued oxidation and prevents the development of conditions under which the Boudouard reaction and spike formation can take place.

INTRODUCTION

Magnesium, alloyed with nominally 0.8 wt% Al and 0.01 wt% Be is known commercially as Magnox Al80 and has been used very successfully for many years as fuel cladding in the nuclear reactors which bear its name. Al is present to improve the mechanical properties of the metal at reactor operating temperatures. The Be addition is needed to increase the oxidation resistance of the metal to an acceptable level in the CO_2 atmosphere of the reactor. Previous studies have shown that the oxidation rate is lowered by many orders of magnitude although the Be concentration is very small. However, there has been no study of the nature of the oxide films produced or of the role of Be in modifying the mechanism of scale formation. The present work was carried out to investigate these aspects of the oxidation reaction. Extreme conditions, 550 C in moist CO_2, well outside the range of normal reactor operating conditions, were employed to speed the reactions and to produce thicker oxides which could be studied more readily by electron microscopical and micro-analytical methods.

EXPERIMENTAL

Four alloys of differing Be contents were supplied for study by the Materials Development Division of the United Kingdom Atomic Energy Research Estblishment at Harwell in the form of sections from semi-finished fuel cans extruded at 400 C. The alloy compositions are given in Table 1. The Be concentrations were determined by British Nuclear Fuels PLC Springfields Laboratories. In all cases the concentrations of other elements present is less than 60ppm except for Zr where the concentration is below 100 ppm. Oxidation kinetics were established using coupons 10 mm square and 1 mm thick cut from the extrudes and mechanically polished to 1 um diamond finish. Standard metallographic and electron optical techniques of examination were employed together with X ray energy dispersive (EDS) and electron energy loss (EELS) methods of chemical analysis. Transverse sections through the metal/oxide interface were prepared for transmission electron microscopy (TEM) as described in (1).

RESULTS AND DISCUSSION

All the alloys exhibited a recrystallised structure with a grain size ranging from 123 um (zero Be) to 163 um (high Be). The oxidation kinetics at 550 C in moist CO_2 are shown in Fig.1 as weight gain per unit area of sample surface versus oxidation time. In all cases, after initial oxide layer formation, there is a period of slow growth, the duration of which decreases with decreasing Be content. The oxide formed is fine grained (60 um) polycrystalline MgO which is dense and continuous. The low rate of growth ($< 7.10^{-5}$ mg.mm^{-2}.hr^{-1}) implies that solid state diffusion, probably along the oxide grain boundaries, is rate limiting at this stage. It is possible that inward diffusion of oxygen (2) rather than the generally accepted outward diffusion of Mg controls this stage of oxide growth. The underlying metal undergoes grain growth during oxidation and boundary migration appears to wrinkle and disrupt the overlying oxide. This eventually breaks down and cracks, presumably as a consequence of the tensile stresses induced by the low Pilling and

TABLE 1
Alloy Compositions

Alloy	Wt.% Al	ppm Be
Zero Be	0.8	<5
Low Be	0.71	19
Medium Be	0.87	78
High Be	0.92	160

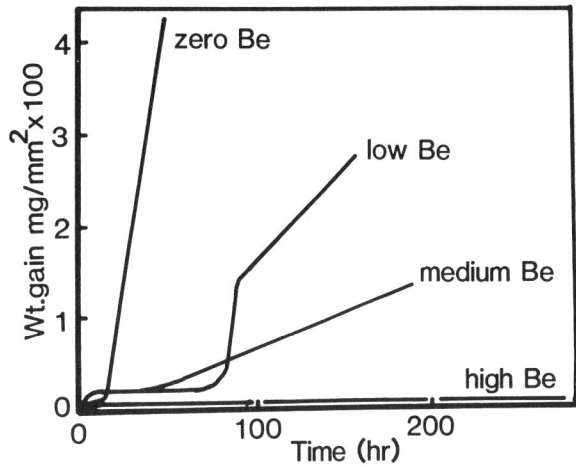

Fig.1   The effect of Be content on the oxidation of Mg 0.8wt% Al in moist CO_2 at 550 C.

TABLE 2
X ray EDS analysis of Al in Zero Be Alloy post Breakaway

Position	At.% Al
Oxide Spike	3.1
Adjacent metal	0
Centre of metal	0.2

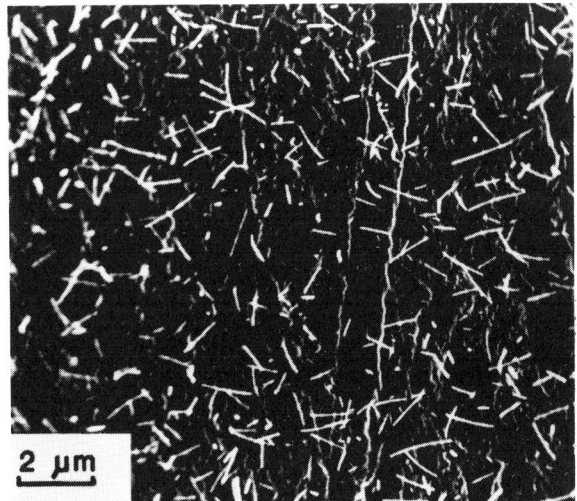

Fig.2   MgO whiskers growing from the rumpled compact oxide on the high Be alloy after 17 hr. oxidation.

TABLE 3
Chemical Analysis of Alloys Pre- and Post Oxidation

Alloy	Oxid.Time(hr)	Al wt.%	Be ppm
Zero Be	0	0.85	<5
"	40	0.69	<5
High Be	0	0.98	165
"	210	0.98	21

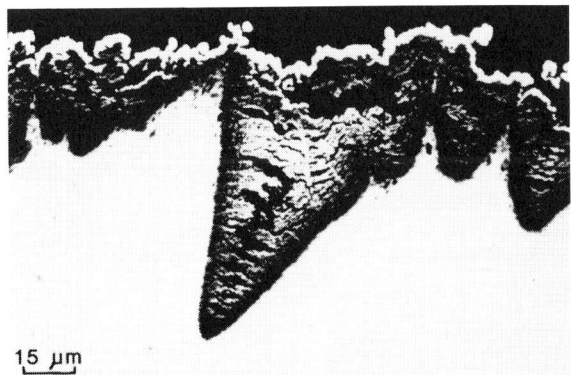

Fig.3   Spikes of porous oxide growing into the zero Be alloy after 46 hr. oxidation.

Bedworth ratio for Mg/MgO (0.81). In the zero Be alloy flocculent oxide outgrowths are produced and spread over the entire sample surface. In situ oxidation studies, by the present authors (3), in a high voltage electron microscope provide direct evidence that this results from the vapour phase oxidation of Mg escaping from the unprotected metal. The linear rate of weight gain, after breakdown of the compact oxide, rises sharply (10^{-3} mg.mm^{-2}.hr^{-1}): calculations indicate that this is fully consistent with a reaction limited by the rate of Mg vaporisation.

With increase in Be content breakdown of the compact oxide is delayed, the linear rate constant after breakaway is reduced, and the production of the flocculent outgrowths diminishes. MgO whisker formation is observed (Fig.2) and in the high Be alloy this is the only product of oxidation observed to grow beyond the initial oxide layer (which is none-the-less observed to be porous by this stage in the reaction).

Be clearly serves to inhibit cracking of the compact scale and to reduce the Mg vapour flux (cf zero and low Be alloys). In the zero and low Be alloys breakaway oxidation is also associated with the formation of spikes of porous oxide which penetrate deep into the metal (Fig.3). This form of attack is not limited to grain boundaries but occurs widely and intragranularly. The spike growth direction was found always to lie within the basal plane and grains with this plane shallowly inclined to the surface did not form spikes. The crystallographic dependence of spike formation cannot simply be associated with crystallographic dependence of oxidation rate (4) which can occur as a result of orientation dependent Mg vaporisation rate: this would result in some grains oxidising faster than others over their entire surfaces. They may originate at cracks in the compact oxide where preferential vaporisation takes place but their continued growth once the entire surface oxide has become flocculent requires an additional driving force. Generally in oxidation reactions the breakdown of a planar front and the development of intrusions into the metal is the result of the preferred oxidation of a solute species (5). EDS analysis of transverse sections through such spikes revealed that in this case the oxide in them is rich in Al and substantial depletion occurs in the underlying metal (Table 2). This observation, although typical, is anomalous. Under the prevailing conditions of oxidation the thermodynamic driving force for Al oxidation is less than that for Mg: the latter should therefore oxidise in preference to Al.

In order to investigate the nature of the reaction in more detail transverse section thin foils were prepared for examination in an analytical TEM. A micrograph showing the metal/oxide interface in the zero Be alloy is presented in Fig.4: the fine grained, porous oxide was identified by EDS and electron diffraction to be MgO. The microscopic oxide interface is highly irregular and pores extend to the metal. At the interface, and within the metal to a depth of about 1 um, a large number of plate shaped crystalline precipitate particles are observed: one such is seen edge on in Fig.5. They are nucleated heterogeneously on subgrain boundaries and Fe/Si rich impurity dispersoids present in the alloys. Analysis of a large number of electron diffraction patterns indicates that the precipitates have hexagonal symmetry

with lattice parameters c = 0.581 nm and a = 0.337 nm. EDS analysis revealed the particles to be rich in Al while EELS proved that they contain C but not O. They are, therefore, Al rich carbides. Combined EDS and EELS analysis is consistent with a composition approximating to $Al_4Mg_2C_3$. The EELS also demonstrated that elemental C is present in the oxide close to the metal/oxide interface. The observed Al depletion in the metal and its enrichment in the oxide spikes is thus associated with internal carburisation, rather than oxidation, just ahead of the metal/oxide interface. As oxidation proceeds the Al rich carbides are oxidised into the porous oxide resulting in the observed Al enrichment in the spikes.

The conditions for carburisation are obtained as a consequence of gaseous diffusion to the metal/oxide interface. CO_2 is reduced and CO concentration rises until the conditions for the Boudouard reaction are established ($pCO_2/pCO < 5.67$):-

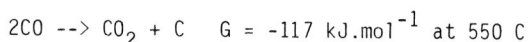

$$2CO \longrightarrow CO_2 + C \quad G = -117 \text{ kJ.mol}^{-1} \text{ at } 550 \text{ C}$$

With the resultant C activity of unity the driving force for elemental Al carburisation is -187 kJ.mol^{-1} whereas that for Mg carburisation is actually positive. Even with a low Al concentration if the activity exceeds 0.001 carburisation can still occur. If Raoult's law is obeyed the activity of Al in the present case is 0.007 at a concentration of 0.8 wt.%.

Clearly Al is more mobile in Mg than C since Al depletion extends over several hundred microns whereas carbides form only within 1 micron of the metal/oxide interface. This provides the conditions needed to destabilise the planar growth front. The outward flux of Al will be proportional to the Al concentration gradient. This steepens if an inward perturbation develops in the oxide front: hence continued growth of the perturbation into a spike is favoured. Any crystallographic anisotropy of Al diffusivity could be reflected in the preferred growth of spikes in directions contained in the basal plane: this cannot be confirmed since there is no data available in the literature concerning such anisotropy.

No Al carburisation is observed in the medium and high Be alloys and thin foils taken through the metal/oxide interface show that a different precipitate phase forms at this position. It consists of tabular crystals of hexagonal cross section (Fig.6): these were identified by electron diffraction to be BeO. The crystals span the interface forming oxide pegs. With increasing oxidation time they grow and, in the high Be alloy, form an effectively continuous layer separating the metal from the outer oxide scale. Clearly the reaction:-

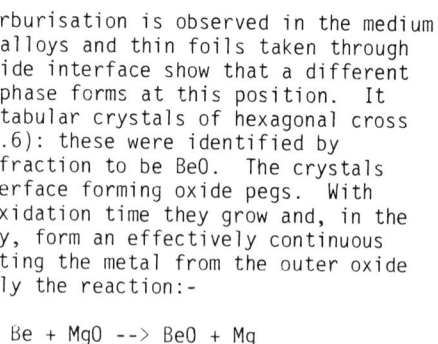

$$Be + MgO \longrightarrow BeO + Mg$$

must take place from left to right under the prevailing conditions. Using standard thermodynamic data the driving force for this reaction at 550 C is calculated to be 15.4 kJ.mol^{-1} if the reactants are all at unit activity. This will not be the case for Be since its concentration is so low. Assuming that the other activities are unity it can be shown that for oxidation of Be to occur its activity must exceed approximately 0.1. If Raoult's law were obeyed the activity of Be in the medium Be alloy would be only about 0.0002. This implies that an

Fig.4   TEM image of the metal/oxide interface
        in the zero Be alloy oxidised for 46 hr.

Fig.5   TEM image of the metal/oxide interface
        in the zero Be alloy. A plate shaped
        Al/Mg carbide is visible edge on across
        the interface.

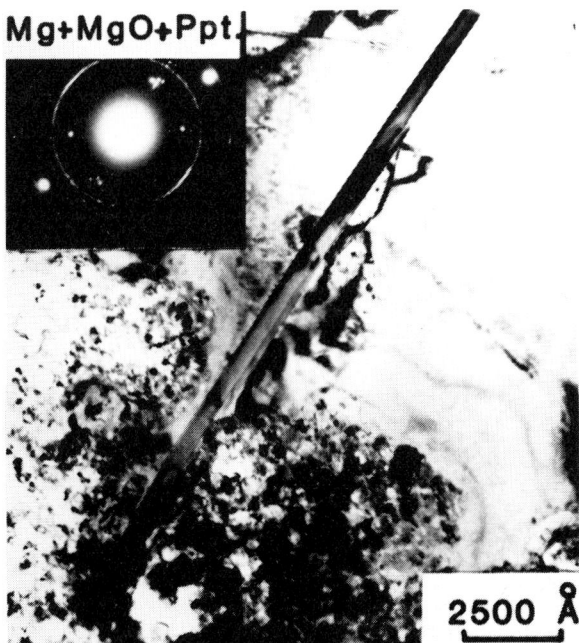

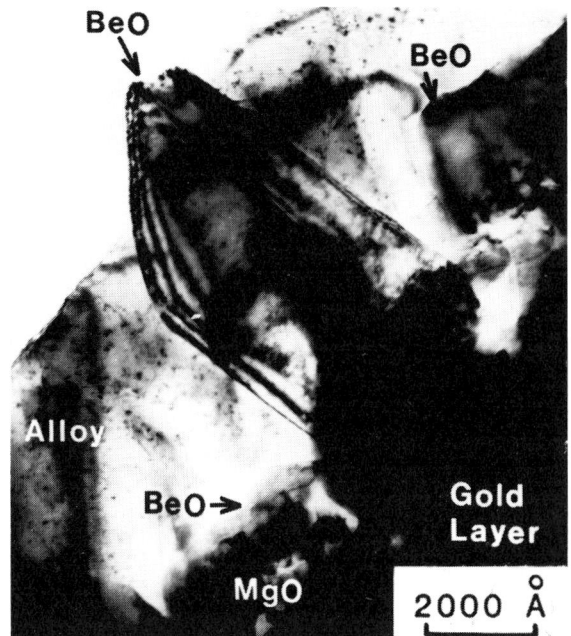

Fig.6   TEM image of the metal/oxide interface
        in the high Be alloy oxidised for 40 hr.

activity coefficient, K, of roughly 550 is
required for Be in Mg: this is a very high value.
However it predicts that the activity should
reach unity at a Be concentration of 0.2 at.%
(0.074 wt.%). This corresponds closely to the
solid solubility limit for Be in Mg established
by Sinelnikov et al (7) at 550 C of < 0.1 wt.%
and thus gives confidence in the calculated value
of K. Using this the activity of Be in the low
Be alloy is found to be only 0.03 which is
considerably lower than the value of 0.1 required
to support preferential Be oxidation. This is
consistent with the lack of protection afforded
by the low Be concentration in this alloy and the
observed absence of BeO.

Kinetically the formation of BeO must be
controlled by the outward migration of Be since
it is found only at the metal/oxide interface and
there is no zone of internal oxidation which
would occur if O diffusion took place into the
metal. The rate of Be diffusion is sufficiently
fast to permit the establishment of an
essentially continuous BeO layer within 17 hours
and it can therefore inhibit further Mg oxidation
before the external compact MgO scale thickens
sufficiently to crack. Even after much longer
times, when the outer oxide becomes granular and
porous the presence of the BeO layer suppresses
Mg oxidation very effectively and the only
additional oxidation observed is the formation of
the MgO whiskers. This might occur at gaps in
the BeO layer through which Mg can migrate but no
experimental evidence for this has been obtained.
The prevention of thickening of the porous MgO
neccessarily inhibits the Boudouard reaction. CO
is carried away from the surface via the gas
stream and does not rise to the required
concentration for carbon deposition.
Consequently there is no driving force for the
breakdown of the planar growth front which is
afforded by the internal carburisation of the Al
and spike formation, so damaging in the zero and
low Be alloys, cannot take place. Chemical
analysis confirms that no Al depletion occurs in
the metal in the alloys protected by the BeO:
correspondingly very substantial Be depletion is
confirmed (Table 3). After very extended
oxidation the Be concentration remaining in the
metal would be too low to protect it from
oxidation if the BeO layer were to be removed or
were to spall off. In this regard it may be
important that the tabular BeO crystals span the
metal/oxide interface and can thus key the oxide
to the metal via their action as pegs: such a
mechanism has been described in other alloy
systems by Whittle and Stringer (8).

CONCLUSIONS

Breakaway oxidation of Mg 0.8 wt.% Al in
moist CO_2 at 550 C is associated with vapour
phase oxidation of Mg and production of a highly
porous oxide.

The Boudouard reaction results in the
internal carburisation of Al and instability of
the oxide growth front. This results in the
growth of deep oxide filled spike shaped
intrusions into the metal.

Be present in greater than a critical
concentration oxidises in preference to Mg and
forms a protective BeO layer below the MgO.
Vapour phase oxidation and the Boudouard reaction
are suppressed giving rise to a marked reduction
in oxidation rate.

ACKNOWLEDGEMENTS.

The receipt of an SERC research studentship by
one of the authors (AJM) and the financial
support of the UKAEA Harwell and BNFL are
gratefully acknowledged together with the
provision of laboratory facilities by
Prof.D.W.Pashley.

REFERENCES

1. A.J.MORRIS. PhD Thesis, University of London
   (1985).

2. H.HASHIMOTO, M.HAMA and S.SHIRASAKI, J.Appl.
   Phys. 43, (1972), 4828.

3. A.J.MORRIS and H.M.FLOWER. Proc. 7th Int.Conf.
   High Voltage Electron Microscopy. Berkeley,
   Calif. U.Calif. (1983), 267.

4. K.SPLICHAL and L.JURKECH. J.Nucl.Mats. 48,
   (1973), 277.

5. N.BIRKS and G.H.MEIER. Intro. High Temp.
   Oxid. Metals. (1983) Edward Arnold. London

6. P.G.SHEWMON. Trans.AIME, 206, (1956), 918.

7. K.D.SINELNIKOV, V.E.IVANOV and V.F.ZELENSK.
   Proc. 2nd. Int. Conf. Peaceful Uses of Atomic
   Energy, 5, (1958), 234.

8. D.P.WHITTLE and J.STRINGER. Phil. Trans. Roy.
   Soc., 295A, (1980), 309.

Phase diagram of liquid
magnesium—aluminium—manganese alloys

B C OBERLÄNDER, C J SIMENSEN,
J SVALESTUEN and A THORVALDSEN

*B C Oberländer and C J Simensen are with the
SI Senter for Industriforskning, Oslo, Norway.
J Svalestuen and A Thorvaldsen are with
Norsk Hydro a.s, Porsgrunn, Norway.*

Introduction

Aluminium and manganese are main elements in
several commercial casting magnesium alloys.
One important effect is that they form
intermetallic phases in molten metal. The main
aim of this microstructure investigation of
ternary magnesium-aluminium-manganese alloys has
been to determine the phase diagram in the
temperature range 660—760° C and for alloy
compositions 0—10 wt% aluminium and 0—1.5 wt%
manganese. The work was carried out by means of
special sampling techniques and analyses of the
intermetallic phases by X-ray diffraction and
microprobe.

Experimental Melt Treatment

A series of magnesium-aluminium melts were
saturated in manganese by addition of $MnCl_2$ at
780° C. The alloys investigated (Table 1) were
then isothermally heat treated at either 750° C,
710° C or 670° C as described in Figure 1. One
hour holding time was selected on the basis of
preliminary work which revealed that the melt
had come close to equilibrium after this treatment.

Analytical methods

Sampling of the melt was carried out after each
isothermal treatment (Fig. 1).
Analysis of the molten materials was carried
out in the following ways:

1 Standard samples were cast in iron crucibles.
The chemical composition of the metal was
determined by emission spectrometer, and the
microstructure was determined by light microscopy,
scanning electron microscopy and microprobe
analysis.

2 Rapidly quenched materials were made by
casting against a spinning wheel. The
concentration of elements in liquid solution
and the composition of the intermetallic phases
were determined by microprobe analysis.

3 Samples were cast, remelted and kept for 15
minutes at the holding temperature. They were
then centrifuged with a velocity of 2000 rpm until
the metal had solidified.(1)
The intermetallic phases which had gathered in a
20 µm bottom layer of the sample, were analyzed by
X-ray diffraction.

Results

The rapidly quenched materials contained regularly
shaped particles, 1—5 µm in diameter as shown in
Figure 2.
The standard cast had in addition dendritically
shaped particles at grain boundaries and cell
boundaries (Fig. 3). These particles were
generally rich in aluminium and were identified
as $Al_{11}Mn_4$, Al_4Mn, and Al_6Mn.
They were frequently found to form heterogeneously
on small intermetallic particles and could
therefore be identified as forming during cooling
to room temperature.

The equilibrium phases in the alloy investigated
were found to be:

1 β-Mn(Al). The particles were cubes with the
concentration manganese-39 at% aluminium. X-ray
diffraction pattern of the phase revealed that
the phase was either cubic with a = 0.641 nm (2)
(with some extra reflections) or tetragonal with
a = b = 0.641 nm and c = 0.660 nm, (Fig. 4).

2 MnAl with 48—52 at% manganese. This phase is
hexagonal with lattice parameter a = 0.269 nm,
c = 0.438 nm (2). The particles are discs.

3 Mn_5Al_8 with a composition close to
stoichiometry. The particles were thin discs with
a hexagonal structure (3).

4 Mn_4Al_{11}-platelets. It was found that
equilibrium phases contained between 0.1 and 0.3
at% silicon.

The solubility of manganese in liquid solution
increased with increasing temperature and with
decreasing aluminium content. The phase diagram
of the magnesium-rich corner was constructed on
the basis of these microprobe measurements of
liquid solution. The diagrams are shown in
Figures 5—7. β-Mn(Al) is the dominant phase
below 2 at% aluminium, AlMn is the equilibrium
phase in 4 wt% aluminium above 700° C and
Al_8Mn_5 is the dominant phase for high aluminium
concentrations and low temperatures. Finally,
Mn_4Al_{11} is present in melts of Mn-10 wt% Al at
low temperatures.

Conclusions

The phase diagram of the low alloyed magnesium-
aluminium-manganese system has been established at
temperatures of between 670 and 750° C.
The solubility of manganese at 700° C decreases
from 2—5 wt% to 0.25 wt% on alloying to 10 wt%
aluminium. The equilibrium intermetallic phase
changes discontinuously from β-Mn to $Al_{11}Mn_4$ by
increasing the aluminium content from 0 to 10 wt%
and by decreasing the temperature from 750° C to
670° C.

REFERENCES:

(1) C J Simensen: Met. Trans. 12 B, 733 (1981)
(2) W Köster and E Wachtel: Z. Met.kde (1960), 261
(3) K Schubert et al: Naturwis. 47, 303 (1960)

Table 1

Chemical composition of the magnesium
alloys investigated.

Alloy no.	Concentration (wt%) of elements							
	Mg	Al	Mn	Fe	Si	Cu	Ni	Zn
92-8	Base	0.8	1.35	0.004	0.01	0.001	0.0014	0.01
92-6	Base	4.0	0.77	0.004	0.01	0.001	0.0007	0.01
92-15	Base	7.2	0.61	0.005	0.01	0.001	0.0010	0.01
92-9	Base	9.8	0.31	0.018	0.01	0.001	0.0008	0.01

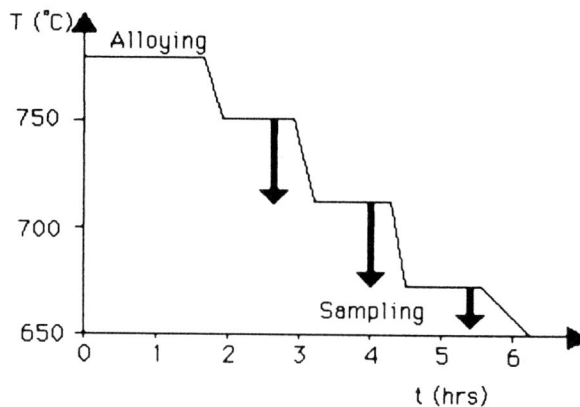

Figure 1

Heat treatment of the magnesium alloys
investigated.
Samples of the melt were taken after each
hour-long isothermal heating.

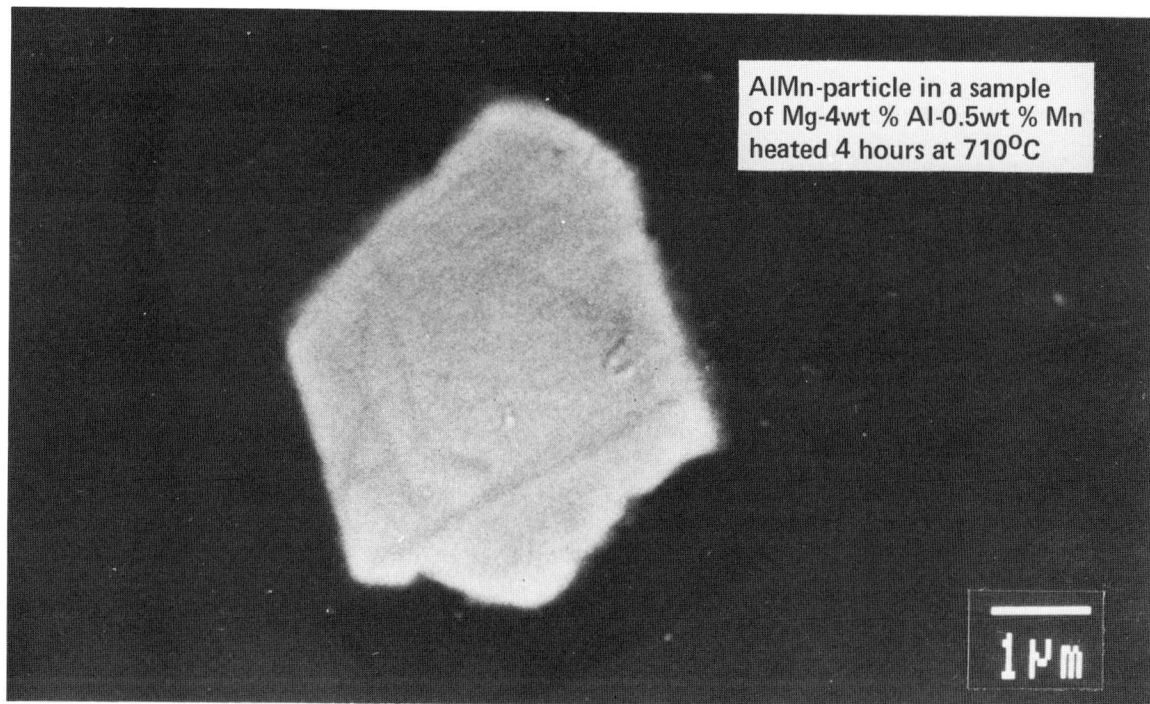

AlMn-particle in a sample
of Mg-4wt % Al-0.5wt % Mn
heated 4 hours at 710°C

Figure 2

135

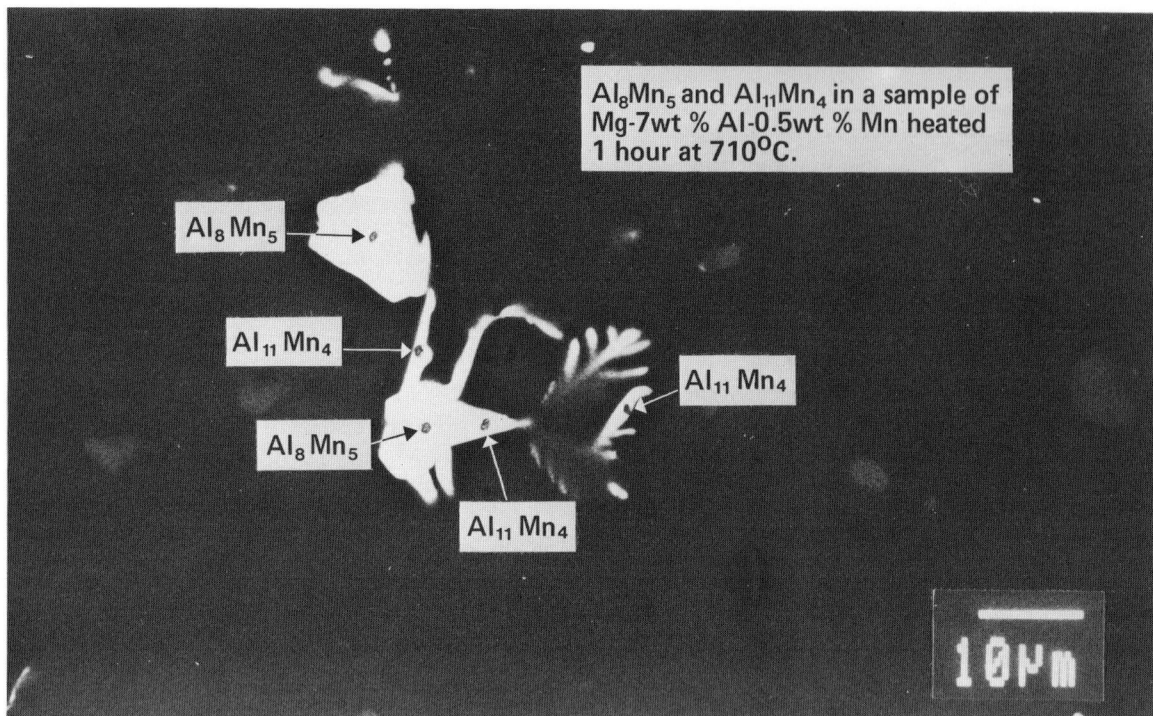

Figure 3

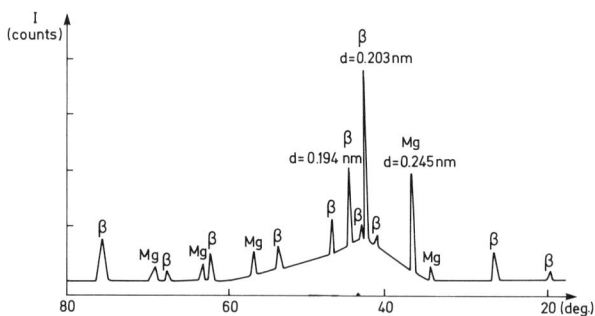

Figure 4

X-ray diffraction pattern of β-Mn(Al) in
centrifuged sample of Mg-0.8 wt% Al 1.1 wt% Mn.

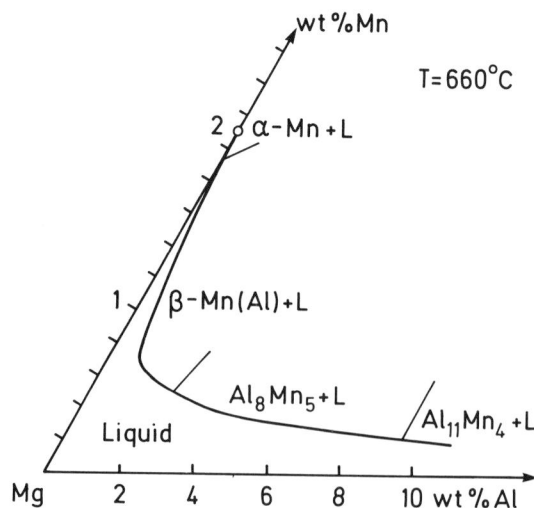

Figure 5

Phase diagram of the magnesium-rich corner
of Mg-Al-Mn at 660° C.

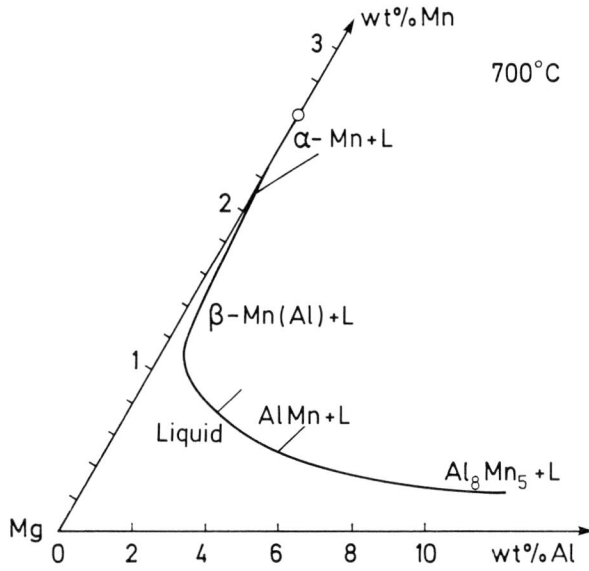

Figure 6

Phase diagram of the magnesium-rich corner
of Mg-Al-Mn at 700° C.

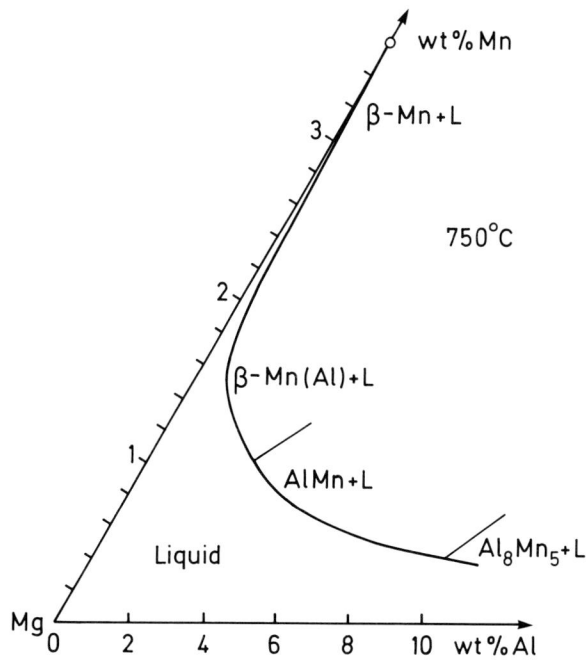

Figure 7

Phase diagram of the magnesium-rich corner
of Mg-Al-Mn at 750° C.

Tensile and creep fracture
of a Mg–Y–RE alloy

H KARIMZADEH, J M WORRALL,
R PILKINGTON and G W LORIMER

*The authors are all with the Department of
Metallurgy and Materials Science, University of
Manchester/UMIST, Manchester, England.*

SYNOPSIS

The tensile and creep fracture characteristics of
a Mg-Y-RE alloy have been examined using optical
and electron optical techniques. Tensile
specimens were aged at 200°C for 170h, 250°C for
20h and 300°C for 8h. Failure during tensile
testing at room temperature was initiated at the
retained eutectic phase located at grain
boundaries after solution treatment. Crack
propagation was either transgranular or
intergranular, depending on the ageing treatment.
Specimens for creep testing were aged at 250°C
for 16h following solution treatment and creep
testing was carried out in the range 250-300°C;
the fracture mode was found to be predominantly
intergranular at most stresses and temperatures.

INTRODUCTION

Magnesium casting alloys which are used for high
temperature (T > 250°C) applications usually
contain about 3wt% thorium, as well as zirconium,
for grain refinement, and possibly additional
alloying elements. While the alloys have good
creep properties, they have only moderate tensile
strengths and thorium is classed as a radioactive
material. Magnesium-yttrium-rare earth alloys
have a higher yield strength than the
thorium-containing alloys and good high
temperature creep properties. In the present
paper the tensile properties at room temperature
and the creep properties in the temperature range
250-300°C have been examined. Particular
reference has been made to the modes of tensile
and creep fracture.

EXPERIMENTAL

The alloy WE54X, of composition Mg-Nd 1.82wt%-Zr
0.52wt%-Misch metal 6.85wt% (75%Y + 25%RE) was
solution treated at 525°C (~20° below the solidus
temperature), for 8 hours and hot water quenched.
Tensile specimens were aged to peak hardness at
200°C (170h), 250°C (20h) or 300°C (8h). Creep
specimens were aged at 250°C for 16h which
corresponded to standard commercial practice.
Tensile testing was performed with a Zwick 1474
machine at the premises of Magnesium Elektron
Ltd. using test pieces of gauge length 32 mm and
diameter 7 mm. Creep tests, using test pieces
with 50 mm gauge length and diameter 6.4 mm, were
carried out under constant load in air at
temperatures of 250°C, 275°C and 300°C at both
Magnesium Elektron and Manchester University.
Specimen temperatures (maintained to ±2°C) were
monitored using Chromel-Alumel thermocouples
attached to the specimen gauge length.

Longitudinal sections of test specimens were
prepared for optical microscopy using standard
metallographic techniques followed by etching in
2% Nital. Fracture surfaces of tensile and creep
specimens were examined using a Philips 505
scanning electron microscope. To examine the
possibility of cavity nucleation during creep
testing, a number of tests were interrupted
before failure and specimens were cooled under
load. They were then removed, cooled to -50°C
and fractured. The fracture surfaces were
examined both optically and electron optically.

RESULTS AND DISCUSSION

The details of the age hardening precipitation
sequence and the crystallography of the
precipitates will be discussed in detail
elsewhere.[1] Briefly, in the peak hardness
condition after ageing at 200 and 250°C the alloy
contained the intermediate body centred
orthorhombic β' precipitates, while following
ageing at 300°C the equilibrium face centred
cubic β phase was formed. Both precipitates
formed as plates on the {10$\bar{1}$0} planes of the
matrix. As well as the strengthening
precipitates in the matrix there was a
significant volume fraction of remnant
intermetallic eutectic phase at the grain
boundaries which remained after solution
treatment at 535°C, Fig. 1.

The room temperature tensile properties of
the alloy after ageing to peak hardness at 200,
250 and 300°C are given in Table 1.

Figures 2a,b show scanning and optical
micrographs of the tensile fracture from the
sample aged at 200°C. These show a mainly
transgranular fracture. When the ageing
temperature was increased to 250°C a mixed mode
fracture occurred: the fracture was approximately
50% transgranular and 50% intergranular. After
ageing at 300°C the fracture was almost
completely intergranular (Figs. 3a,b). The
transgranular portions of the fractures were
ductile in character. The retained eutectic

phase played an important role in the initiation of tensile failure in all specimens, Fig. 1. An apparently anomalous result has been obtained for the sample aged at 300°C: the fracture is almost completely intergranular (and brittle) while the ductility, as measured by the elongation of tensile specimens, is greater than in specimens with mixed or predominantly transgranular (and ductile) failure, Table 1. This type of observation has been previously interpreted[2] to suggest that the increase in elongation implies increased resistance to crack initiation, whilst the transition to a brittle intergranular failure mode implies that crack propagation resistance is markedly reduced.

Table 2 gives details of creep tests carried out at 250°C, 275°C and 300°C. Examination of the relationship between the secondary creep rate ($\dot{\epsilon}_s$) and the applied stress (σ), on the basis of a power law expression of the form $\dot{\epsilon}_s = A\sigma^n \exp(-Q/KT)$, where A is an arbitrary constant, n is the stress exponent, and Q is the apparent activation energy for creep produces a stress exponent ~4. Calculations using the modulus corrected stress[3] suggest that the apparent activation energy is ~230 kJ.mol^{-1}.

Examination of specimens after test revealed the eutectic phase frequently to be associated with the creep fracture process (Fig. 1). A typical optical micrograph showing the distribution of creep damage after failure is shown in Fig. 4. It will be seen that the cracking was quite coarse and at the centre of grain boundary facets rather than being associated with grain boundary triple points. Creep cracking has frequently been reported to be associated with triple point wedge cracking[4] and in the present material, this is clearly not the case. Creep fracture in most metals and alloys is usually a consequence of cavity nucleation and growth[5]; observations on the present failed specimens revealed no evidence of cavitation. Several creep tests were therefore interrupted before failure and broken open at low temperature to expose damaged facets. No traces of cavitation damage were detected. It is suggested that these observations are evidence for a decohesion mechanism of creep fracture in the present material.

It is also appropriate to consider one further metallographic observation with respect to the development of the microstructure during test. During prior ageing, a metastable β' precipitate was created which transformed during test into the equilibrium β phase[1]. A precipitate free zone was observed to develop during testing adjacent to grain boundaries which were perpendicular to the tensile axis, Fig. 5. This observation is taken to indicate that Herring-Nabarro creep is a possible operative mechanism[6].

CONCLUDING REMARKS

Attention is drawn to two important features in this paper: an increase in ageing temperature of a Mg-RE alloy of the present type produces a change in the tensile fracture mode from transgranular to intergranular while at the same time the material elongation increases; secondly, it is unusual for cavity nucleation and growth to be absent under creep conditions.

ACKNOWLEDGMENT

This work has been supported by Magnesium Elektron Ltd.

REFERENCES

1. G.W. Lorimer, this Conference, pp.

2. R. Pilkington, G. Willoughby and J. Barford, 1971, Met.Sci., 5, pp. 1-8.

3. S.S. Vagarali and T.G.Langdon, 1981, Acta Met., 29, pp. 1969-82.

4. J.A. Williams, 1967, Acta Met., 15, pp.1119-1124.

5. H.E. Evans, Mechanisms of Creep Fracture, 1984, Elsevier, pp. 1-24.

6. J.P. Poirier, 1985, Creep of Crystals, Cambridge, pp. 195-202.

Table 1

Tensile Data as a function of Ageing Treatment.

Treatment	U.T.S.. (MPa)	ϵ_f (%)
170h at 200°C	285	0.6
33h at 250°C	304	2
8h at 300°C	263	6

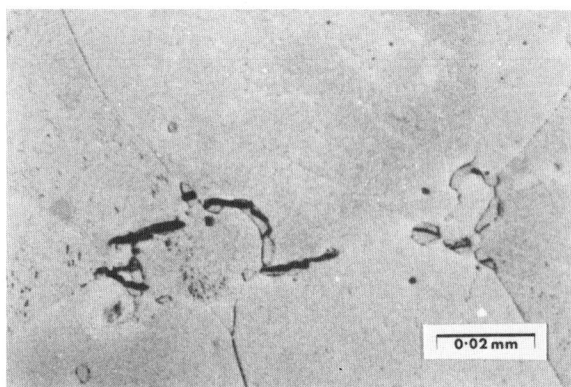

Table II

Creep Data

Temperature (°C)	Stress (MPa)	$\overset{\circ}{\epsilon}_{min}$ (h)	t_f (h)	ϵ_f %
300	93.5	4.8×10^{-3}	6.7	10.50
300	63.4	8.6×10^{-4}	39	11.50
300	45.2	1.8×10^{-4}	130	4.80
300	44.8	1.8×10^{-4}	210	5.40
300	31.2	5.4×10^{-5}	250*	3.20
300	31.2	4.8×10^{-5}	439	6.20
275	93.5	4.8×10^{-4}	48	4.50
275	62.5	8.9×10^{-5}	236	4.50
275	44.5	2.2×10^{-5}	544	3.00
275	31.5	7.3×10^{-6}	1200*	1.20
250	125.0	2.1×10^{-4}	70	3.00
250	93.3	5.2×10^{-5}	333	3.33
250	62.6	1.3×10^{-5}	1008	2.60

* Test interrupted before failure.

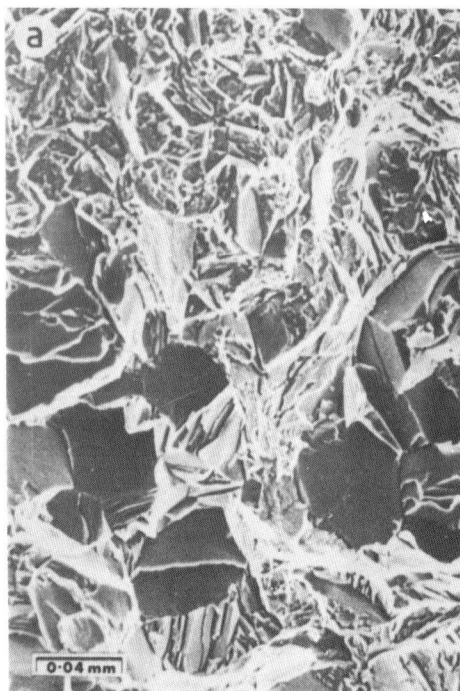

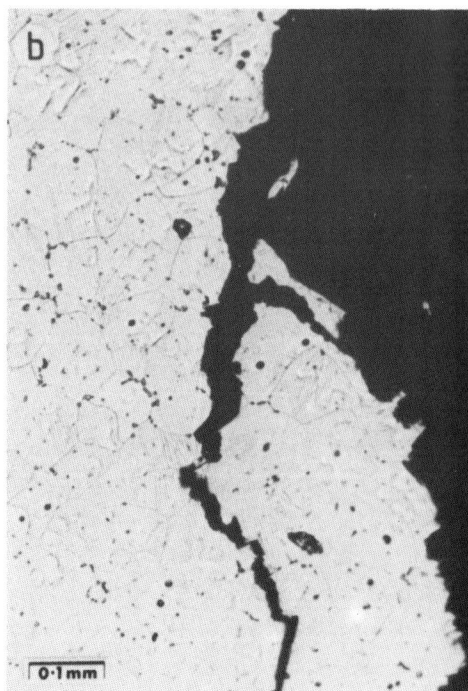

Fig. 1. Optical micrograph showing intermetallic eutectic phase and associated cracking at grain boundaries.

Fig. 2   a. Scanning electron micrograph, and
          b. Optical micrograph.
Fracture surface of tensile specimen aged for 170h at 200°C.

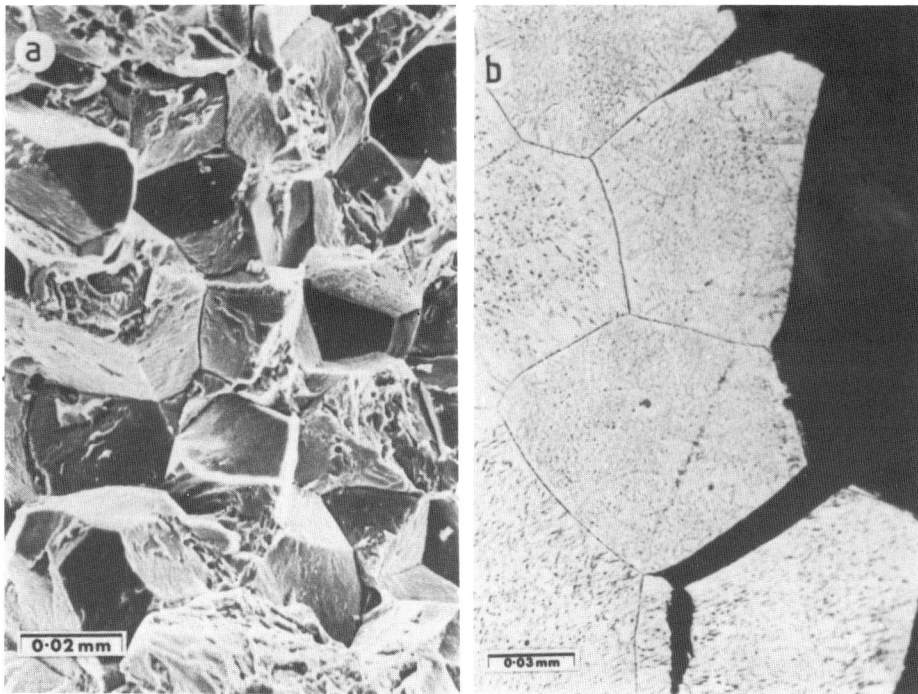

Fig. 3 a.   Scanning electron micrograph and
      b.   Optical micrograph.
Fracture surface of tensile specimen aged for 6h
at 300°C showing intergranular fracture.

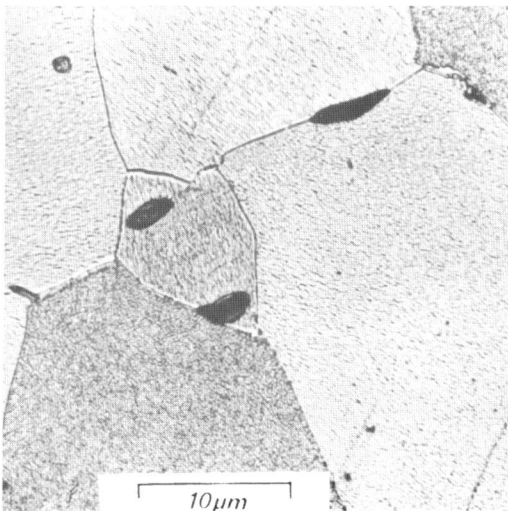

Fig. 4.   Optical micrograph showing distribution
of creep cracking.   (Specimen tested at 45 MPa,
300°C, t_f ~ 130h).

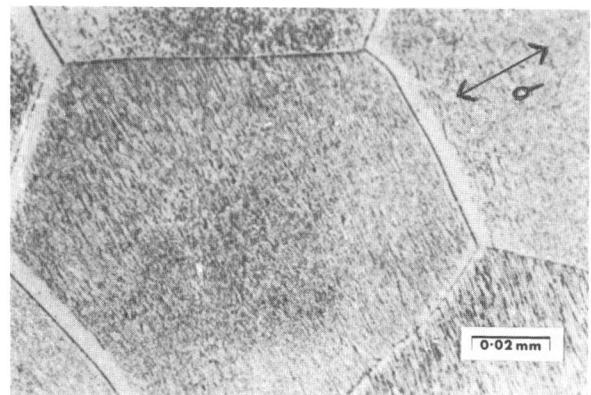

Fig. 5.   Optical micrograph of the development of
precipitate free zones on the transverse grain
boundaries of creep specimens.   (Specimen tested
at 60 MPa, 250°C, t_f ~ 3000h).

AUTHOR INDEX